Bryant Research

The Agentic Enterprise

A Leader's Guide to Navigating the AI Revolution

Steven Bryant and Kevin Moos

Bryant Research

The Agentic Enterprise

A Leader's Guide to Navigating the AI Revolution

Steven Bryant and Kevin Moos

The Agentic Enterprise: A Leader's Guide to Navigating the AI Revolution

By Steven Bryant and Kevin Moos

Printed in the United States.

Published by Bryant Research Corporation, 369-B 3rd Street, Suite 366, San Rafael, CA 94901

Bryant Research Corporation books may be purchased for educational, business, or sales promotional use. A PDF edition is distributed freely by the authors and is available at https://BryantResearch.AI

April 2025 First Edition

Revision History

04302025:	First Release
05232025:	Release 1.3 (.rev10)

ISBN: 978-0-9962409-4-9 (Paperback Edition)

Table of Contents

Preface: Navigating the Agentic Enterprise

The Strategic Imperative: Leading in an Age of Intelligent Action

You are leading at a pivotal moment of technological acceleration, an era where the very fabric of business is being rewoven by Artificial Intelligence. The rapid ascent of capable AI agents – intelligent systems designed to perceive, reason, learn, and act with increasing autonomy – is not merely another incremental trend. It is a fundamental force actively reshaping competitive landscapes, operational logic, customer expectations, and the nature of work itself. These are not just advanced tools; they represent a profound leap beyond prior automation waves, introducing new levels of cognitive capability and operational agency into your enterprise.

While you may not have chosen this inflection point, navigating its inherent ambiguity to secure strategic advantage, mitigate profound risks, and uphold ethical stewardship is now a core leadership mandate. In this dynamic environment, delay is no longer a viable strategy; it is an active concession in a rapidly evolving market, an abdication of the responsibility to shape a future where technology serves human values and enhances enterprise performance. The Agentic Enterprise is emerging, and its contours will be defined by the leaders who act decisively and wisely today.

Core Insights from This Guide: A Blueprint for Principled Transformation

This guide cuts through the pervasive noise and hype surrounding Agentic AI. It offers pragmatic, experience-based perspectives specifically for C-suite decision-makers, operating executives, and strategists tasked with navigating this complex transformation. Our analysis, informed by real-world challenges and successes, reveals critical insights demanding immediate and sustained executive attention:

- **A Fundamental Capability Shift Demands New Leadership:** AI agents are "outcome pursuers," a distinct evolution from the 'instruction followers' of earlier automation like Robotic Process Automation (RPA). This shift fundamentally alters operational possibilities, redefines competitive dynamics, and necessitates a new playbook for strategy, governance, and leadership.
- **The Urgency is Real, The Stakes Are Escalating:** Clear market signals – accelerating adoption, massive strategic investments, and a reshaping of talent demands – confirm that the agentic shift is moving decisively from experimental labs into core business operations. This creates a 'burning platform' where inaction incurs compounding strategic costs and cedes ground to more agile, intelligent competitors.
- **Trust is the Non-Negotiable Foundation for Scale and Success:** Realizing the transformative potential of agentic systems at scale is impossible without deep, verifiable trust. This requires proactively embedding ethical principles, robust security, and operational transparency through ***Trusted AI by Design*™**. This comprehensive approach to governance, woven into the entire AI lifecycle from inception, is not a constraint but an enabler of confident, responsible innovation.
- **The Human Platform is Your Enduring Differentiator:** Paradoxically, intelligent automation elevates, not diminishes, the strategic value of unique human capabilities: sophisticated judgment, frame-breaking creativity, genuine empathy, and unwavering ethical oversight. Cultivating this human platform – through intentional cultural transformation, strategic workforce design, and a commitment to Visibility, Opportunity, and Fair & Relevant Evaluation – is critical for unlocking sustainable advantage and ensuring AI augments human potential equitably.

Your Leadership Roadmap: From Understanding to Enduring Advantage

Our objective in this guide is twofold: first, to provide strategic clarity on the nature and implications of the Agentic Enterprise; and second, to equip you with the operational frameworks and governance principles needed to build and lead one effectively, responsibly, and with enduring impact. This guide maintains a consistent leadership vantage point, translating technological complexities and ethical nuances into actionable insights. The chapters ahead provide a structured path to:

- **Demystify Core Concepts (Chapter 1 & 2):** Understand the strategic imperative of the Agentic Enterprise, and gain foundational clarity on AI agent capabilities, types, limitations, and the unique risks they present.
- **Implement for Impact (Chapter 3):** Learn to execute strategic catalyst pilots, design for scalability, and manage the AI agent lifecycle with rigor and foresight to translate vision into tangible value.
- **Architect Trust (Chapter 4):** Master the principles and practices of ***Trusted AI by Design*™**, establishing robust governance, risk management, and ethical frameworks as the bedrock of your agentic transformation.
- **Cultivate the Human Platform (Chapter 5):** Discover how to foster the right culture, develop essential 'agent fluencies', redesign roles for human-AI synergy, and champion ethics and fairness to unlock the full potential of your workforce.
- **Sustain Advantage (Chapter 6):** Ensure long-term strategic alignment, anticipate future evolutions in AI agency, and embed the disciplines required to future-proof your enterprise for enduring resilience and leadership.

Consider this your comprehensive briefing on navigating what's next – a guide to transforming uncertainty into opportunity, risk into resilience, and technological potential into principled, sustainable performance. It is your roadmap for confidently leading your organization into, and through, the agentic future. Let's begin.

Your Guides to the Agentic Enterprise

Steven Bryant is a pioneering voice in the strategic implementation of Artificial Intelligence within complex organizations. As a founder of Bryant Research, he draws upon three decades of executive leadership at global institutions like Accenture and Wells Fargo, where he spearheaded transformative enterprise AI and automation initiatives that redefined operational efficiency and customer engagement. Steven's unique synthesis of profound, hands-on transformation experience and deep technical AI fluency – grounded in an MS in Computer Science (AI) from Georgia Tech, where he is a College of Computing faculty member, and an MBA from the University of San Diego – positions him as a preeminent advisor guiding today's leaders through the multifaceted challenges and immense opportunities of the Agentic Enterprise.

Kevin Moos is a distinguished authority in architecting trust and embedding data privacy as foundational pillars of enterprise growth and innovation. With over three decades of experience advising senior leaders at premier firms including Primitive Logic and Accenture, Kevin has been at the forefront of navigating the intricate global landscape of privacy regulations (GDPR, CCPA, and beyond). His expertise in translating complex regulatory mandates into practical, scalable, and business-enabling solutions is integral to the successful implementation of the *Trusted AI by Design*™ framework – an indispensable approach for organizations deploying AI agents responsibly. Kevin holds a BS in Mathematics and Computer Science from Stanford University and is a sought-after strategist for building the trustworthy Agentic Enterprises of tomorrow.

Chapter 1: The Agentic Enterprise

The Defining Moment: Beyond Automation to True Agency

Leadership is perpetually tested by defining technological shifts – inflection points where strategic choices dictate future relevance and market dominance. We stand at such a threshold today, propelled by the rapid, undeniable ascent of Artificial Intelligence (AI). While the capabilities of Generative AI and the fluency of Large Language Models (LLMs) capture headlines and spark widespread imagination, the critical evolution driving fundamental business transformation – the one demanding immediate, decisive executive focus – is the emergence of AI agents.

Consider an AI system not merely analyzing market trends but autonomously identifying underserved niches, drafting initial outreach campaigns, and dynamically allocating marketing budgets across channels based on real-time performance – all while continuously learning and adapting its strategies. This is the domain of AI agents. These systems, built upon sophisticated foundational models, are engineered not just to communicate, analyze, or automate discrete tasks, but crucially, to act with a significant degree of delegated authority and purposeful intent. This represents a profound leap beyond previous technological waves, which largely centered on enhancing processing power, expanding connectivity, or refining rule-based automation.

The current transition centers on cognition: the capacity for these AI agents to understand complex goals, reason through multifaceted options, learn from dynamic interactions (with humans, data, and other systems), and execute intricate multi-step tasks with an unprecedented level of dynamic autonomy. This capability doesn't just incrementally enhance existing automation; it fundamentally reshapes the landscape of decision-making, the architecture of value creation, and the very nature of human-machine collaboration. The executive imperative is therefore clear and urgent: to re-architect the organization for a new operational reality – the era of the Agentic Enterprise. This is not merely about adopting a new tool; it is about embracing and leading a new paradigm.

Why AI Agents Demand Executive Attention Now

AI agents – intelligent systems capable of perceiving their environment, reasoning through options, acting to achieve goals, and collaborating with dynamic autonomy – represent a quantum leap beyond traditional automation and even sophisticated analytics. This advance demands unwavering strategic focus, even from organizations already leveraging advanced data analytics, Robotic Process Automation (RPA), or foundational LLMs for content generation or analysis. Prior tools primarily assist human analysis, automate discrete, explicitly instructed tasks, or generate outputs based on prompts. Agents, however, introduce the capacity for autonomous, goal-directed action and orchestration, enabling entirely new operating models and fundamentally challenging existing competitive dynamics.

They integrate and synthesize capabilities that rewrite the rules of market engagement and operational efficiency:

Autonomous Execution & Orchestration

- **Beyond RPA**: While RPA excels at automating repetitive, rule-based tasks by mimicking human keystrokes (e.g., data entry from one system to another), AI agents operate on a different plane. They can initiate and complete complex, multi-step workflows, often applying judgment learned from data and contextual understanding, without constant human intervention. For example, an RPA bot might process invoices according to a fixed script. An AI agent, conversely, could manage the entire accounts payable ecosystem: ingesting invoices in myriad formats, validating them against purchase orders and delivery receipts, identifying discrepancies, autonomously communicating with vendors to resolve issues, seeking necessary approvals based on learned financial thresholds, and scheduling payments – all while adapting to new invoice formats or evolving compliance policies.
- **Strategic Implication for Leaders**: This signifies a shift from task automation to end-to-end process automation and, ultimately, to outcome orchestration. The

critical questions become: *Which of our core complex processes, previously demanding significant human coordination and nuanced judgment, can now be strategically reimagined with AI agents at their core? How do we effectively govern systems that not only execute but also make critical operational decisions within defined strategic boundaries?*

Adaptive Reasoning & Problem Solving

- **Beyond Predefined Logic**: Traditional analytics might predict outcomes based on historical data, and rule-based systems rigidly follow predefined logic paths. AI agents, however, can evaluate complex options, navigate ambiguity with greater efficacy, and dynamically plan multi-step paths to achieve strategic goals, even when confronted with novel or unforeseen situations. For instance, a supply chain agent, detecting a significant disruption at a key global port, wouldn't merely flag the issue. It could autonomously evaluate alternative shipping routes, consider intricate cost, time, and risk trade-offs, negotiate with alternative carriers via APIs in real-time, and re-plan logistics to minimize impact – all while learning from the success or failure of its adaptations to inform future responses.

- **Strategic Implication for Leaders**: This capability allows businesses to architect significantly more resilient and adaptive operations. Leaders must ask: *How can agentic reasoning fundamentally enhance our organization's ability to respond to market volatility, geopolitical instability, or unexpected operational disruptions? What level of decision-making authority can we confidently and responsibly delegate to agents capable of sophisticated, adaptive problem-solving, and what new forms of intelligent oversight are required?*

Contextual Memory & Personalized Interaction

- **Beyond Stateless Chatbots**: Basic LLMs and traditional chatbots often lack persistent memory of past interactions, treating each query as a new, isolated event. AI agents, conversely, can retain and meaningfully apply organizational knowledge, individual user preferences, comprehensive interaction histories, and evolving contextual understanding across multiple engagements. A customer service agent, for example, wouldn't just answer an isolated query; it would recall the customer's entire service history, understand their previous issues and resolutions, anticipate their current needs based on recent activity or product usage, and provide highly personalized, proactive solutions or recommendations, leading to a far more coherent, efficient, and satisfying experience.

- **Strategic Implication for Leaders**: This enables hyper-personalization at scale, transforming customer and employee engagement from transactional to relational. Key questions emerge: *How can we strategically leverage agent memory to create truly differentiated and continuously improving stakeholder experiences that build lasting loyalty? What are the critical data governance, privacy (e.g., GDPR, CCPA), and cybersecurity implications of agents maintaining rich contextual memory, and how do we manage these complex responsibilities proactively and ethically?*

Collaborative Integration & Human Augmentation

- **Beyond Isolated Tools**: AI agents are engineered to work alongside human teams and coordinate seamlessly with other systems and agents. They can augment human judgment by providing timely insights, automating complex sub-tasks within a larger collaborative effort, or orchestrating intricate workflows that span multiple human and digital actors. For example, in a complex strategic research project, an agent could gather and synthesize information from diverse global sources, identify relevant internal and external experts, facilitate collaborative sessions by managing schedules and documents, and even draft initial summaries and analyses, thereby allowing human researchers to focus on higher-level critical thinking, strategic interpretation, and breakthrough innovation.

- **Strategic Implication for Leaders**: This redefines the nature of teamwork and the strategic role of human expertise. Leaders need to explore: *How can AI agents best augment our most valuable human talent, freeing them from cognitive drudgery for higher-value strategic work? What new team structures, collaborative processes, and skillsets are needed to maximize human-agent synergy and drive innovation? How do we ensure seamless handoffs, maintain clear accountability, and foster mutual trust in these hybrid workflows?*

This is not incremental improvement; it is a fundamental transformation in operational logic, value creation, and competitive capability. Organizations that understand, embrace, and strategically begin to harness these integrated agentic capabilities gain an unparalleled opportunity to redefine market expectations, create novel value propositions, and establish new, defensible competitive benchmarks. Viewing agents as merely "AI version 2.0" or a more advanced iteration of existing automation fundamentally misunderstands the strategic magnitude of the shift at hand and risks ceding significant, potentially irreversible, ground to competitors who are already architecting entirely new ways to create and deliver value.

"In 2024, U.S. private AI investment grew to $109.1 billion – nearly 12 times China's $9.3 billion and 24 times the U.K's $4.5 billion. Generative AI saw particularly strong momentum, attracting $33.9 billion globally in private investment – an 18.7% increase from 2023." – The State of AI 2025

The Leadership Mandate: Act Now or Be Acted Upon

The strategic conversation surrounding enterprise technology has irrevocably shifted. As prominent research underscores, "A technology wave becomes a board-level issue only when it threatens core economics. Autonomous agents have crossed that line." What began for many organizations as cautious AI experimentation or isolated departmental pilots is rapidly escalating into a core strategic imperative, centered on intelligent software designed not just to analyze or predict, but to act with purpose, intelligence, and delegated authority.

Autonomous agents are no longer theoretical constructs confined to research papers; they are operational realities impacting the bottom line now. They are reading and interpreting complex regulatory and "documents, reconciling vast and disparate datasets across enterprise systems, drafting initial compliant decisions in highly regulated industries, and intelligently escalating exceptions that require nuanced human judgment and ethical consideration. Early deployments are showcasing staggering results: loan adjudication processes, once taking days, are now slashed to mere hours, significantly reducing operational costs while dramatically improving customer experience and access; service ticket volumes are being substantially reduced through intelligent self-resolution capabilities and proactive issue identification before customers are even impacted; significant duplicate spending and operational inefficiencies are being rapidly identified and remediated within complex global procurement workflows. These tangible economic shifts are not only validating initial investments but are also driving exponential growth in enterprise agent spending, with CFOs and boards increasingly aligning capital allocation toward building and expanding these transformative agentic capabilities.

This accelerating momentum signals unequivocally that the inflection point has passed. Automation, supercharged by AI agents, is becoming the primary engine of a competitive reset, simultaneously reshaping customer expectations for speed, personalization, and proactive service; redefining achievable operating margins and efficiency frontiers; and fundamentally altering talent strategy, workforce design, and the very nature of human work. The leadership mandate is therefore unavoidable and urgent. Success in this new era requires embracing and championing four core tenets:

See the Strategic Frontier: Recognize Agents as Programmable Colleagues, Not Just Tools

- **Elaboration**: This demands a fundamental mindset shift from viewing agents as mere "tools" or software enhancements to recognizing them as a new class of "programmable colleagues" or digital team members. These agents can dramatically shorten insight-to-execution cycles by taking on complex cognitive tasks and decision-making processes previously requiring extensive human intervention. This involves a deep understanding of their capabilities to not just automate but to reason, learn, adapt, and act within clearly defined strategic and ethical boundaries.

- **Leadership Action**: Champion a clear and compelling strategic vision where AI agents augment human capabilities across the entire value chain, from R&D to customer engagement. Empower and incentivize cross-functional teams to identify high-impact areas where agentic capabilities can unlock new levels of performance, create entirely new service offerings, or solve previously intractable business problems.

Move Before the Baseline Shifts: Embrace Decisive, Informed, and Agile Action

- **Elaboration**: History consistently demonstrates that decisive movers gain disproportionate advantages during periods of significant technological transformation. Hesitation, prolonged debate, or an overly cautious "wait and see" approach invites obsolescence as agent-native competitors or fast-adopting incumbents rapidly redefine performance benchmarks, cost structures, and customer expectations. The baseline for 'good enough' is shifting at an unprecedented pace.

- **Leadership Action**: Foster an organizational culture of informed risk-taking, rapid learning, and agile experimentation. Personally sponsor and resource well-defined catalyst pilots (as will be detailed in Chapter 3) designed to build internal capabilities, demonstrate tangible value quickly, and learn by doing in a controlled manner. Prioritize speed in learning and strategic adaptation over achieving theoretical perfection in initial deployments.

Govern for Trust at Scale: Embed Responsibility, Ethics, and Safety from Day One

- **Elaboration**: The immense power and autonomy of AI agents necessitate a profound and unwavering commitment to trust, safety, and ethical conduct.

Embedding transparency, auditability, fairness, security, and robust human oversight into agent design, development, and operational protocols from the very beginning is not an optional add-on or a compliance checkbox; it is an absolute executive imperative. This is the foundational philosophy of *Trusted AI by Design*™ (which will be detailed in Chapter 4).

- **Leadership Action**: Mandate the establishment of clear, robust, and adaptable governance frameworks, ethical guidelines, and risk management protocols before scaling agent deployments across the enterprise. Ensure that accountability for agent actions is unambiguously defined and that sophisticated systems are in place to continuously monitor, audit, and control agent behavior within ethical and operational guardrails.

Redeploy Liberated Capacity: Strategically Reinvest and Elevate Human Potential

- **Elaboration**: AI agents will inevitably automate significant portions of existing knowledge work, often freeing up 20-30% (or potentially much more) of human capacity currently consumed by routine cognitive tasks, manual data reconciliation, information gathering, or standardized communications. This liberated human potential is an immensely valuable strategic asset.
- **Leadership Action**: Develop proactive, forward-looking strategies for upskilling, reskilling, and redeploying the workforce. Focus on redirecting human talent towards higher-value activities that AI agents cannot perform – such as complex strategic problem-solving, breakthrough innovation, empathetic customer relationship building, ethical deliberation, and sophisticated oversight of AI systems. Frame this transition as a strategic opportunity to elevate human contribution and create more meaningful work, not merely as a mechanism to reduce headcount.

Figure 1.1 Leadership Pillars

Organizations that internalize these principles and begin architecting for advanced agentic capabilities are not just preparing for a distant future; they are actively shaping it. They are positioning themselves to set the definitive performance benchmarks, define new market categories, and lead the competitive dynamics for the coming decade and beyond.

The leadership mandate requires moving decisively before the competitive baseline shifts irreversibly, architecting advantage through focused execution and embedded trust.

Signals You Cannot Afford to Ignore: The Market's Verdict is In

This urgency is not speculative; it is grounded in concrete, quantifiable market signals that paint an undeniable picture of rapid transformation. Tracking shifts in enterprise adoption patterns, strategic capital allocation priorities, and evolving talent demands provides a clear and unambiguous radar for discerning strategic inflection points. Ignoring these signals, or dismissing them as temporary hype, risks fundamentally misinterpreting the strategic landscape and the breathtaking speed of its evolution. Three critical indicators demand immediate and sustained executive attention:

Enterprise Adoption Has Tipped from Experimentation to Full-Scale Operationalization

- **The Signal:** Over 70% of large enterprises now report running live, impactful agentic workflows deeply embedded in core business functions – such as customer service operations, financial reconciliation and reporting, and dynamic supply chain management. This represents a twofold increase in adoption reported just 18 months prior. Furthermore, the nature of these deployments has decisively shifted from isolated, small-scale proofs-of-concept to integrated, enterprise-grade solutions handling significant operational loads and directly impacting key financial and operational performance indicators.
- **The Implication for Leaders:** AI, particularly in its agentic form, has decisively moved from the R&D lab and innovation outposts to the very heart of business operations. Your peers, your most aggressive competitors, and even your key customers are actively operationalizing these capabilities to gain efficiency, enhance experiences, and create new value. The question is no longer *if* agentic AI will impact your industry, but *how quickly, how profoundly,* and whether you will lead or be led by this transformation.

Budgets Follow Strategy: Capital is Flowing Decisively and Massively to Agentic Capabilities

- **The Signal:** Initiatives focused on developing and deploying sophisticated AI agents are consistently securing top-level investment approval and significant budgetary allocations, mirroring the enterprise-wide strategic prioritization previously seen with pivotal shifts like cloud adoption and enterprise-wide data analytics platforms. Decisive capital is flowing – with an estimated $4.6 billion invested globally into agent-powered applications and platforms in 2024 alone, an astounding eightfold increase year-over-year. This investment is increasingly directed towards building agentic capacity as a core, differentiating, and enduring business engine, not just for isolated point solutions or incremental improvements.
- **The Implication for Leaders:** The financial markets, venture capital, and internal corporate capital allocation processes are unequivocally recognizing the transformative potential of agentic AI and are placing substantial, long-term bets

on its success. Organizations that are slow to strategically allocate resources and invest in building these capabilities risk falling critically behind in innovation, efficiency, and market responsiveness.

Talent Market Realigns: New Competencies Signal New Frontiers of Value Creation

- **The Signal:** Job postings demanding specific, advanced skills related to the design, development, deployment, and governance of AI agents – such as “““ ,” “autonomous systems management,” “AI ethics and governance,” and “human-agent collaboration design” – have surged sevenfold in the past year alone. This confirms that these capabilities are no longer niche technical skills confined to specialized teams but are rapidly becoming critical operational assets that generate tangible economic value and are fundamentally reshaping workforce competencies and organizational structures.

- **The Implication for Leaders:** The skills required to build, manage, govern, and effectively work alongside AI agents are becoming mainstream requirements for competitive success. Organizations must urgently adapt their talent acquisition strategies, internal development programs, and employee retention initiatives to secure and cultivate these new competencies, or risk facing a significant and debilitating skills gap that severely impedes their agentic transformation and growth.

Dismissing these converging macro pulses – rapid and deep adoption, decisive and large-scale capital allocation, and fundamental talent market realignment – is akin to ignoring the overwhelming early data on the rise of the mobile internet or the enterprise-wide shift to cloud computing. History offers stark and unforgiving lessons about the consequences for those organizations that waited too long to recognize and act upon such fundamental, market-altering shifts. The data is clear: the agentic revolution is not on the horizon; it has arrived and is accelerating.

“AI business usage is also accelerating: 78% of organizations reported using AI in 2024, up from 55% the year before.” – The State of AI 2025

The Burning Platform: Why Delay Is Now a Rapid Value-Leak Strategy

Market signals confirm the agentic shift is well underway and accelerating. But why the profound, almost existential, urgency for leaders to act? Beyond aggregate trends and market statistics, tangible operational and competitive pressures are intensifying daily, creating a strategic "burning platform" for nearly every organization across all sectors. Maintaining legacy operating models, often reliant on outdated manual processes, siloed data, or brittle first-generation automation, is no longer a safe or prudent option; on the contrary, delay actively leaks value, erodes competitive positioning, and confers compounding strategic disadvantage at an alarming rate.

This intense pressure manifests across multiple critical fronts, driven by rapidly changing operational economics and stakeholder expectations that are being fundamentally and permanently reset by the demonstrated capabilities of intelligent, autonomous systems:

Customer Tolerance for Friction Has Utterly Collapsed

- **The Pressure:** In an environment where agent-native competitors or digitally transformed incumbents can offer instant, hyper-personalized, proactive service and seamless experiences, traditional friction points – slow response times, cumbersome interfaces, inconsistent information across channels, reactive problem-solving – are becoming entirely unacceptable. Customers now benchmark their experiences against the best-in-class service that AI agents enable, regardless of industry. Patience has evaporated.
- **The Value Leak:** Any friction, delay, or impersonality in your customer journey now rapidly erodes brand affinity, directly drives customer churn, and significantly diminishes customer lifetime value. Delay in deploying AI agents to streamline processes, personalize interactions, and proactively address needs means you are actively frustrating your customers and, in effect, pushing them towards more agile, responsive, and intelligent competitors.

Human Capacity Represents the Wrong, and Most Expensive, Bottleneck for Cognitive Work

- **The Pressure:** Knowledge workers, your most valuable and often most expensive human capital, still spend an unsustainable portion of their time – often estimated at 30-50% or more – on low-value, repetitive "swivel-chair" cognitive work. This includes manually reconciling data across disparate systems, re-keying information between applications, searching for basic information buried in complex knowledge bases, or managing routine, standardized communications. This is a profound misallocation of talent, a significant drag on productivity, and a major impediment to innovation and strategic thinking.

- **The Value Leak:** AI agents are specifically designed and increasingly adept at eliminating this type of cognitive inefficiency. They can slash cycle times, drastically reduce error rates, improve data quality and consistency, and liberate human intellect, with these gains flowing directly and immediately to the P&L. Relying solely on constrained and expensive human capacity for automatable cognitive tasks is becoming economically untenable and deeply demoralizing for your workforce. Delay here means perpetuating significant operational inefficiency and chronically underutilizing your organization's most critical asset: its people.

Scale Is Decoupled from Headcount, Permanently Rewriting Competitive Dynamics

- **The Pressure:** Lean, agent-powered entrants and agile, digitally transformed incumbents can now launch new products, bill for complex services, market with precision, and support global customer bases faster, more efficiently, and with far greater personalization than organizations burdened by traditional, labor-intensive back offices and rigid front-office processes. Technology, specifically AI agents, allows intelligence, operational reach, and sophisticated customer engagement to scale almost infinitely, independent of linear hiring growth.
- **The Value Leak:** Market share, customer loyalty, and ultimately, profitability will increasingly follow agent-amplified speed, responsiveness, personalization, and agility – not just organizational size, historical market position, or brand legacy. Delay in leveraging AI agents for intelligent scale means you are competing with an inherently less efficient, less adaptable, and less responsive operating model, making it progressively harder to grow profitably, enter new markets, or react effectively to evolving customer needs and competitive threats.

Regulatory Scrutiny and Compliance Demands Intensify for All Automated Decisions

- **The Pressure:** Global regulations, such as the EU AI Act, and a growing array of sector-specific mandates (e.g., in finance, healthcare, transportation) are increasingly requiring auditable, transparent, fair, secure, and risk-managed AI systems, especially those involved in significant decision-making processes or interacting directly with consumers and their data. Designing for compliance, security, and responsible governance from the very inception of AI initiatives is rapidly becoming a baseline expectation and a non-negotiable requirement for market access and societal license to operate.

- **The Value Leak:** Waiting to address these complex requirements until late in the development cycle, or treating them as an afterthought, risks severe operational disruption, costly and time-consuming remediation sprints, significant fines and penalties, and potentially irreversible reputational damage simply to meet baseline legal and ethical standards. Delay in embedding *Trusted AI by Design*™ principles into your agentic initiatives means accumulating substantial technical, ethical, and regulatory debt that will inevitably have to be paid, often at a much higher cost.

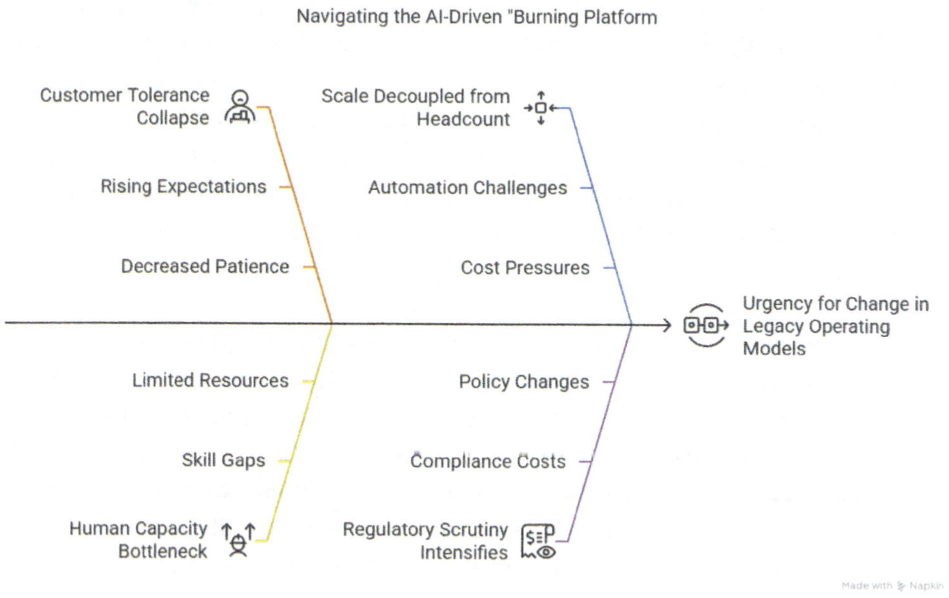

Figure 1-2. The Burning Platform

Together, these powerful and converging forces – radically shifting customer demands, persistent internal operational inefficiencies, new competitive dynamics driven by scalable intelligence, and an evolving, more stringent regulatory landscape – create an undeniable and intensifying burning platform. Every manual handoff, every delayed customer response, every spreadsheet-driven analytical workflow, every missed opportunity for data-driven personalization becomes a self-imposed tax on your organization's performance and agility, effectively funding faster, more intelligent, and more adaptable competitors. Early adopters who are strategically leveraging AI agents are not just optimizing existing processes; they are fundamentally resetting the performance baseline against which your organization will be judged – by customers, by investors, by regulators, and by current and future talent.

Overwhelming signals – rapid adoption, decisive capital allocation, talent market realignment – confirm the agentic shift has crossed into core business operations.

The Cost of Inaction: Watching Your Strategic Advantage Disappear

Understanding the burning platform reveals *why* immediate action is necessary; understanding the true cost of inaction reveals the steep, compounding, and potentially irreversible *price* of delay. A "wait and see" stance, or tentative, under-resourced experimentation in the face of this transformative shift, is not prudent risk management; it is an active – and often unacknowledged – decision that directly subsidizes competitors, progressively erodes shareholder value, and sets the organization on a dangerous path toward strategic irrelevance. Failing to engage decisively and strategically with agentic AI doesn't just leave your organization behind; it makes catching up exponentially more difficult, if not impossible, as the negative consequences of inertia compound relentlessly.

Consider the tangible, escalating costs of inaction:

- **The Runaway Efficiency Dividend Accruing to Competitors:** Organizations that embrace agentic capabilities early and effectively are not just cutting costs; they are fundamentally re-architecting their operational models and cost structures, creating a runaway efficiency dividend. They establish compounding learning loops, where AI agents continuously optimize processes, enhance data quality, automate complex decision-making, and drastically reduce the marginal costs of operation and service delivery. This creates significant, widening cost-structure gaps that become incredibly difficult and prohibitively expensive for slower-moving organizations to close.
- **The Imposition of a Market-Multiple Penalty and Diminished Investor Confidence:** Financial markets and astute investors recognize this divergence in operational capability, strategic foresight, and future readiness. A discernible market-multiple penalty is increasingly being imposed on organizations perceived as laggards in AI adoption and agentic transformation. Studies consistently confirm that AI-forward firms – those demonstrating a clear strategic commitment and tangible, impactful results from AI – achieve higher growth

rates, improved profitability, and superior valuation multiples. Tentative, delayed, or ineffective adoption translates directly into lower shareholder value, reduced access to capital, and diminished investor confidence.

- **A Talent Exodus, Not Merely Slow Erosion:** Top performers – especially in technical, analytical, strategic, and creative roles – actively seek environments where intelligent tools amplify their impact, automate drudgery, and allow them to focus on challenging, meaningful, and innovative work. When agile competitors begin eliminating tedious cognitive work with AI agents, redesigning roles for human-agent collaboration, and offering more dynamic, AI-augmented career paths, your best and brightest talent will inevitably migrate to organizations that better value their skills, invest in their growth, and provide superior opportunities for impact. This is not a slow, manageable erosion of talent; it can quickly become a rapid and damaging talent exodus, actively pushing your most valuable human capital – and their future innovations – toward your keenest competitors.

- **A Shrinking Strategic Canvas and Escalating Future Costs:** Operational lagging, a direct consequence of inaction, significantly impacts strategic agility and future options. Key strategic moves – such as new market entries, mergers and acquisitions, the launch of innovative products and services, or responses to disruptive threats – increasingly assume an agent-level operational baseline for speed, efficiency, data-driven decision-making, and customer engagement. Lacking this foundational agentic capability means every new strategic initiative carries inherently higher execution costs, longer payback periods, greater operational risk, and a reduced probability of success, effectively throttling the range of viable strategic options available to your organization and limiting its ability to adapt and grow.

- **Real-Time Brand Dilution and Accelerated Customer Churn:** Customer expectations are not static; they are dynamic and constantly reset to the best, most seamless, and most personalized experience encountered *anywhere*, regardless of industry. If your organization's processes involve delays, inconsistencies, impersonal interactions, or a lack of proactive support while competitors are leveraging AI agents to resolve issues instantly, provide anticipatory service, and deliver hyper-personalized experiences at scale, your brand suffers tangible damage in real time. Customer satisfaction plummets, frustration grows, negative sentiment spreads virally, and churn to more agile alternatives inevitably follows. Operational lagging in the agentic era directly translates into a declining brand experience, diminished customer loyalty, and a steady loss of market share.

History offers blunt and cautionary precedents for organizations that misjudged the speed, scope, and impact of major technological shifts: retailers slow to embrace e-commerce ceded vast, irretrievable ground to Amazon and other digital natives; Blockbuster ignored the disruptive potential of streaming and ultimately collapsed; Kodak, despite inventing the first digital camera, hesitated on fully embracing digital photography and lost its century-long market dominance. In the agentic era, delay is not a neutral stance; it is an active opportunity subsidy paid directly to faster, more adaptive, and more intelligent movers in your market. The cost of inaction is not just stagnation; it is a managed, or often unmanaged, decline into strategic irrelevance.

Delay actively subsidizes competitors, erodes market valuation, pushes top talent away, and shrinks your strategic options.

The Leadership Opportunity: Architecting Enduring Advantage in the Agentic Shift

While the costs of inaction are undeniably severe, this period of profound technological disruption also opens a significant and time-sensitive window for decisive, visionary, and courageous leadership. Major technological shifts invariably reward leaders and organizations that can accurately perceive the emerging landscape, courageously translate potential threats into scalable, sustainable advantages, and effectively mobilize their organizations to act before the market fully adjusts and new competitive moats are firmly established. Autonomous AI agents, with their transformative capabilities, represent precisely such an opportunity – a chance not just to adapt, but to architect enduring competitive advantage and redefine industry leadership.

Encouragingly, the journey into the Agentic Enterprise can, and often should, begin surgically and strategically, unlike some past large-scale transformations that required massive, upfront capital expenditure on physical infrastructure or enterprise-wide system replacements. A single, well-chosen catalyst pilot (as will be detailed in Chapter 3) – one that is meticulously framed strategically, resourced appropriately, and sharply focused on a tangible, P&L-visible business problem or a critical stakeholder pain point – can deliver rapid, measurable, and highly visible proof points of value. This initial success does far more than just validate the technology; it ignites a powerful "proof-of-future" flywheel within the organization. It fundamentally alters perspectives on what's possible, tangibly demonstrates that AI agents can be governed effectively, ethically, and safely, builds crucial stakeholder confidence and buy-in, and begins to attract the internal and external talent essential for broader, sustainable transformation. A pilot that demonstrably pays for itself, or delivers clear and compelling strategic value quickly, builds immense credibility and unleashes vital organizational momentum.

Seizing this advantage requires intentional, focused, and unwavering action from the highest levels of leadership:

Reframe Vision & Sponsor Focused Execution: Moving Decisively from Tools to True Transformation

- **Elevate the Strategic Conversation:** Move the internal organizational dialogue decisively beyond merely adopting "AI tools" or implementing "automation software" to embracing the strategic imperative of "operating model redesign," "cognitive augmentation," and "agentic transformation." Clearly and consistently articulate how AI agents will enable fundamentally new ways of creating value, serving customers, empowering employees, and achieving strategic objectives.
- **Anchor All Initiatives to Core Value Levers:** Ensure that all initial and subsequent agent initiatives are explicitly and demonstrably anchored to core business value levers – such as radically improving customer experience and loyalty, achieving a step-change reduction in cost-to-serve, dramatically accelerating product innovation cycles and speed-to-market, or significantly enhancing operational resilience and risk management. This ensures strategic focus, facilitates clear measurement of true impact, and justifies investment.
- **Provide Decisive Executive Air Cover and Resources for a Catalyst Pilot:** Personally sponsor, visibly champion, and provide unwavering executive support and dedicated resources for a high-impact catalyst pilot. Target a P&L-visible workflow, a critical operational bottleneck, or a significant stakeholder pain point where success will be undeniable and broadly recognized. Mandate a tight, aggressive timeframe (e.g., 60-90 days) for delivering a Minimum Viable Agent (MVA) that produces measurable, impactful results. Ensure shared, ambitious KPIs across all involved business, technology, and operations teams, and critically, mandate Minimum Viable Governance (including clear ownership, comprehensive risk assessment, and robust ethical review) from the absolute outset to generate a credible, responsible, and replicable win.

Build Unshakeable Confidence: Making Trust a Prerequisite for Scale

- **Embed *Trusted AI by Design*™ as a Non-Negotiable Foundation:** Recognize that sustainable opportunity can only be seized if the underlying AI systems can be unequivocally trusted by all stakeholders. From day one of any agent initiative, however small, mandate the deep integration of robust Data Privacy, comprehensive Risk Management, and proactive Cybersecurity protocols into the agent's architecture, development process, and operational lifecycle (as will be detailed in Chapter 4). This is not an afterthought or a compliance hurdle; it is a foundational requirement for success and societal license to operate.
- **Design for Intrinsic Transparency, Auditability, and Human Oversight:** Ensure that transparency into agent decision-making processes, clear pathways for effective human oversight and intervention, and comprehensive, immutable audit logging are architected as core, indispensable features of any agent deployment, not as optional extras or bolt-ons. This builds confidence, enables effective governance and accountability, and facilitates rapid learning and adaptation.
- **Treat Security, Privacy, and Compliance as Co-Architected Enablers, Not Impediments:** Engage your organization's security, privacy, legal, and compliance teams as integral, strategic partners in the design, development, and deployment process from the earliest stages. Their expertise is crucial for proactively navigating complex risks, ensuring adherence to evolving regulations, and building resilient, trustworthy systems that can withstand scrutiny.
- **Codify Every Lesson into a Living, Reusable Enterprise Playbook:** Critically, establish a disciplined process to capture every lesson learned – both successes and failures, technical insights and cultural shifts – from initial pilots and subsequent deployments. Document effective design patterns, proven governance protocols, successful risk mitigation strategies, user feedback, and performance metrics into a dynamic, continuously updated, and easily accessible enterprise-wide playbook. This accelerates future deployments, reduces redundant effort and costly mistakes, builds institutional knowledge and capability, and fosters a culture of continuous improvement.

Converting the pervasive hype surrounding AI agents into tangible, sustainable, and defensible competitive advantage requires tracking specific leading indicators of progress and maturity. Positive signals that confirm your organization is architecting advantage responsibly, ethically, and securely include: your comprehensive agent strategy featuring prominently and confidently in executive communications, board discussions, and investor earnings calls; your internal audit function constructively referencing and validating your AI governance frameworks and operational controls in their reviews; top talent (both internal and external) actively seeking to be involved in your organization's agent initiatives due to their ambition and responsible approach; and new agent-driven projects demonstrably leveraging established patterns (e.g., achieving ≥60% reuse of codified best practices, standardized integration connectors, or proven governance protocols) from your organization's evolving agentic playbook. These are clear signs that trust is being systematically built, capabilities are maturing rapidly, and strategic momentum is accelerating across the enterprise.

Case Study: Regional Bank Confronts the Future

Scenario: Apex Regional Bank, facing declining mortgage market share and rising customer complaints about slow processing (CX impact), realized its operational costs were unsustainable. Fintech competitors offered loan approvals in days, highlighting Apex's inefficiency. This wasn't just an operational issue; it became a board-level concern as the core economics were threatened. Leadership recognized the strategic imperative: inaction meant obsolescence.

Agentic Solution Connection: The leadership mandate was clear – they needed to move beyond incremental improvements. Apex initiated a strategic assessment to understand how AI agents, fundamentally different from their existing automation, could reshape their mortgage origination process. This involved exploring agents capable of autonomous execution and adaptive reasoning for tasks like initial document validation, representing their crucial first step in architecting an Agentic Enterprise to avoid the compounding cost of inaction.

Key Questions for Leaders

Strategic Understanding & Unified Vision: How will we ensure our entire leadership team, including the Board, shares a common, deep, and strategically aligned understanding of AI agents as "outcome pursuers" with transformative cognitive capabilities? How do we move beyond the technical hype to focus collective attention on their potential to fundamentally reshape our operating model, value proposition, and competitive landscape? What is our shared, ambitious, yet realistic vision for the Agentic Enterprise within our specific industry context and competitive environment?

Quantifying the Risk of Inaction & Assessing the Competitive Baseline: Considering the clear market signals and the compounding costs of inaction outlined, what is the specific, quantifiable strategic and financial risk (e.g., in terms of lost market share, eroded margins, diminished innovation capacity, talent attrition) to our business if we delay significant, strategic agentic adoption by another 12-18 months? How are our most agile competitors and new market entrants leveraging agentic capabilities, and how is that rapidly shifting the baseline for performance, customer expectations, and cost structures in our industry?

Identifying High-Impact Catalyst Opportunities & Defining Success: Which specific core value levers (e.g., radically improving customer acquisition and lifetime value, achieving a step-change reduction in enterprise-wide cost-to-serve, dramatically accelerating innovation cycles and speed-to-market for new offerings, significantly enhancing operational resilience and proactive risk mitigation) present the most compelling initial opportunities for high-impact, strategically aligned agentic transformation within our organization? What P&L-visible workflow or critical stakeholder pain point could serve as our first 60-90 day catalyst pilot, and what would constitute undeniable success for that pilot?

Embedding Trust and Responsible AI from Inception: What immediate, concrete, and non-negotiable steps will we take as a leadership team to begin embedding *Trusted AI by Design*™ principles – including transparency, accountability, fairness, security, privacy, and robust human oversight – into our earliest agent explorations, proofs-of-concept, and pilot projects? How will we ensure that governance, ethics, and risk management are treated as foundational enablers of innovation and speed, rather than as afterthoughts or bureaucratic obstacles? How will we guarantee our first pilots are not just technically successful but also demonstrably trustworthy and ethically sound?

Committing to Strategic Resource Allocation & Long-Term Capability Building: Are we, as a leadership team and an organization, truly prepared to allocate the necessary strategic capital, dedicate our best talent (including significant investment in upskilling existing teams and acquiring new, specialized competencies), and provide unwavering executive sponsorship required to move beyond isolated experiments to a programmatic, sustained, and enterprise-wide agentic transformation? What is our multi-year plan for building enduring internal capabilities for advanced agent design, sophisticated deployment, adaptive governance, and continuous innovation in the agentic space?

Conclusion: Embarking on the Agentic Journey with Clarity, Courage, and Purpose

The evidence converges, and the message for leaders is unequivocal: the Agentic Enterprise is not a distant future scenario but the new, rapidly solidifying center of gravity for value creation, demanding immediate, informed, and decisive leadership focus. The market signals are flashing with increasing intensity, the penalties for inertia are significant, compounding, and potentially existential, and the window for architecting durable, differentiating competitive advantage is open – but it will not remain so indefinitely. Your strategic decisions and bold actions in the coming months will significantly influence whether your organization successfully leads this transformation, merely struggles to follow, or is ultimately left behind by more agile and intelligent competitors.

This guide is designed to be your trusted strategic companion on this critical journey, offering leadership frameworks, actionable insights, and practical guidance grounded in real-world experience to foster clarity, enable effective and ethical governance, and drive decisive, impactful action. The path to becoming a successful, resilient, and leading Agentic Enterprise requires a deep and nuanced understanding of the technology's transformative potential and its inherent limitations (Chapter 2). It demands a disciplined, strategic approach to execution, scaling, and continuous improvement (Chapter 3). It rests upon an unwavering, foundational commitment to embedding *Trusted AI by Design*™ principles into every facet of your operations (Chapter 4). It necessitates a profound and ongoing focus on cultivating the human platform through new skills, adaptive roles, and truly collaborative human-agent models (Chapter 5). And it requires a resilient, forward-looking strategy for ensuring sustained strategic alignment and proactively future-proofing your organization against continuous technological and market change (Chapter 6).

Welcome to the profound challenge and the immense opportunity of The Agentic Enterprise. The future is not something to be passively awaited; it is something to be actively, intelligently, and responsibly built. The time for bold leadership and decisive action is now.

Chapter 2: Demystifying AI Agents

Beyond the Buzzwords: Foundational Clarity for Strategic Leadership

Chapter 1 established the undeniable strategic imperative: the Agentic Enterprise is not a distant forecast but an emerging operational reality demanding immediate and decisive leadership attention. Successfully navigating this profound transformation requires moving beyond pervasive hype and superficial understanding. To lead effectively in this new era, executives require a foundational grasp of what AI agents truly are, how they function at a conceptual level, their current capabilities, and, critically, their inherent limitations and associated strategic risks.

Many leaders, understandably, grapple with precise definitions and distinctions. Is an AI agent merely a more sophisticated chatbot? How does it fundamentally differ from Robotic Process Automation (RPA) or existing enterprise analytics platforms? Where does advanced automation demonstrably end, and genuine operational autonomy begin? It is vital for strategic decision-making and effective governance to recognize that today's AI agents, while remarkably powerful and rapidly evolving, are not sentient, infallible, or capable of replicating the full spectrum of nuanced human judgment, complex ethical reasoning, or deep contextual understanding. They are sophisticated, powerful tools requiring careful strategic guidance, robust and adaptive governance, and continuous, intelligent oversight from the highest levels of the organization.

This chapter translates essential technical concepts into actionable strategic insights. We will demystify the core components that grant AI agents their unique capabilities, their dynamic operational rhythm, and the key archetypes of AI agents emerging in the enterprise landscape. The objective is to illuminate both their transformative potential and their current boundaries, enabling informed leadership. This isn't about transforming executives into AI technologists; it's about equipping leaders with the critical strategic fluency required to lead confidently, make informed investment and risk management decisions, and architect sustainable competitive advantage in an increasingly agent-augmented world.

From Instruction Followers to Outcome Pursuers: The Fundamental Agentic Leap

Strategic fluency in the agentic domain begins with a clear and decisive differentiation of AI agents from prior generations of automation tools, such as Robotic Process Automation (RPA) and scripted or basic Large Language Model (LLM)-powered chatbots. RPA excels at mimicking human keystrokes to automate highly structured, repetitive, and rule-based tasks but is notoriously brittle. Even minor changes in application interfaces or process steps can often break the automation, requiring costly rework and ongoing maintenance. Traditional chatbots, and even many first-generation LLM applications, typically operate within narrow conversational scripts, lack persistent, rich contextual memory, and struggle significantly with ambiguity, novel situations, or deviations from their training data. Both RPA and these simpler conversational tools are fundamentally instruction followers; they execute predefined steps or respond to explicit prompts within relatively rigid, pre-programmed constraints.

AI agents, however, represent a distinct and far more powerful evolutionary leap: they are engineered as outcome pursuers. Building on significant and accelerating advances in LLMs, machine learning, knowledge representation, and sophisticated planning algorithms, AI agents don't just execute rote steps or generate text based on immediate input. Instead, they leverage a suite of integrated cognitive capabilities to understand, plan for, and autonomously achieve specified, often complex, business goals. They frequently navigate dynamic, unpredictable, and data-rich environments with a degree of intelligent adaptability.

Key distinctions that define this agentic leap and demand leadership attention include:

- **Strategic Goal Orientation & Intent Understanding:** AI agents move beyond the explicit programming required by RPA. Instead of meticulously scripted steps, they can be given higher-level strategic objectives – such as resolving a high-value customer's complex billing issue comprehensively to ensure continued loyalty, or optimizing inventory levels for a product line across global distribution centers to meet forecasted demand while minimizing costs. The agent then autonomously determines the necessary actions, information gathering, and cross-system interactions to achieve this outcome. For leaders, this fundamentally shifts the paradigm from *detailed task definition* to *clear outcome definition and strategic intent articulation*. The focus must be on clearly articulating desired end-states and strategic priorities, empowering agents (within strict and verifiable governance frameworks) to intelligently determine the "how." This necessitates fostering a new level of organizational trust and establishing a robust, auditable framework for delegating operational authority and decision rights.

- **Advanced Reasoning, Planning & Adaptive Problem-Solving:** Unlike systems reliant on simple rules and static logic, AI agents can evaluate multiple, often conflicting, options, handle significant ambiguity and incomplete information with greater sophistication, and make nuanced judgments based on learned patterns or complex business rules. They dynamically plan (and critically, re-plan) multi-step operational sequences, adapting their approach intelligently if conditions change, new information emerges, or initial attempts encounter unforeseen obstacles. This capability enables the design of far more resilient, adaptive, and intelligent enterprise operations. However, it also means agent decision pathways can be less predictable. Therefore, leaders must ensure governance frameworks are designed to effectively oversee, validate, and (where necessary) challenge agent reasoning processes, especially when agents deviate from standard procedures to solve novel problems or respond to unexpected events.

- **Rich Contextual Memory & Continuous, Governed Learning:** Moving beyond stateless interactions and batch updates, AI agents can maintain and strategically utilize a persistent, evolving memory of past interactions, individual user preferences, relevant organizational knowledge, detailed process histories, and critical environmental context. This enables deeply personalized stakeholder interactions, adaptive and consistent behavior over time, and the effective management of complex, long-running processes. While exceptionally powerful for personalization, efficiency, and continuous improvement, this capability raises critical strategic considerations for leaders regarding data privacy (e.g., stringent adherence to GDPR, CCPA, HIPAA), data security, information lifecycle management, data accuracy, and the risk of "context drift" – where an agent might rely on outdated, biased, or irrelevant information. Robust, proactive data governance and ethical oversight are paramount and non-negotiable.

- **Dynamic Tool Use & Cross-System Orchestration:** Agents are not confined to siloed actions or manual integrations. They can be equipped to dynamically select, securely access, and intelligently utilize a wide array of digital tools and enterprise systems – internal and external APIs, core databases, CRM and ERP platforms, communication systems, curated external knowledge sources, and even other specialized AI services. Through these tools, agents gather information, execute actions, and orchestrate complex workflows across the organizational fabric, translating "thought" and "decision" into tangible business operations and measurable real-world impact. For leaders, granting tool use is a high-value, high-risk strategic proposition. It requires stringent, uncompromising governance closely mirroring (and in some critical aspects, necessarily exceeding) the controls applied to human employees with similar levels of system access and operational authority. This includes clearly defined and narrowly scoped permissions (adhering rigorously to the principle of least privilege), secure multi-factor authentication and authorization mechanisms, comprehensive and immutable audit trails, and potentially, the enforced separation of duties for critical transactions.

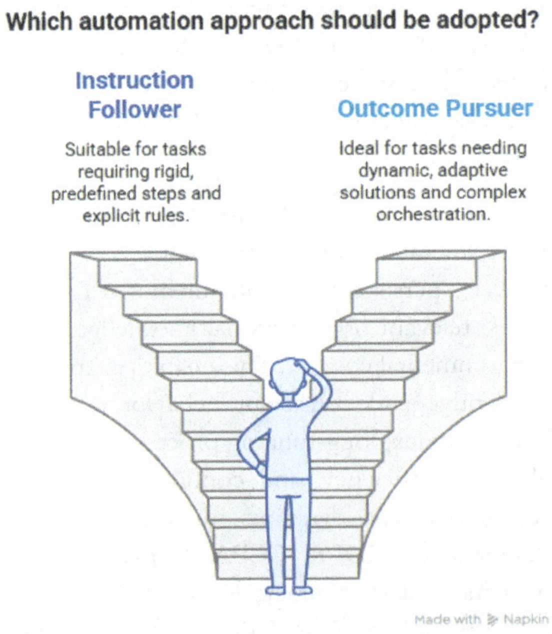

Figure 2-1. Automation Approach Choice

Where RPA automates discrete, often mundane, tasks and chatbots handle relatively simple, frequently informational, queries, AI agents can orchestrate complex, end-to-end business processes and manage multifaceted strategic outcomes. They effectively extend human intent and intellect with a significant degree of delegated judgment and operational autonomy. This leap represents a profound discontinuity in technological capability and operational potential.

Today's hyper-competitive global environment demands unprecedented speed, deep and dynamic hyper-personalization, continuous adaptation to shifting market conditions, and sophisticated cross-functional orchestration – capabilities inherently limited in static, rule-based automation paradigms. AI agents enable a strategic shift from merely automating existing tasks more efficiently to fundamentally re-architecting *how work gets done and how value is created.* This involves moving from optimizing routines to intelligently delegating complex cognitive work and orchestrating dynamic, adaptive, value-creating outcomes. Recognizing, embracing, and strategically capitalizing on this shift demands a leadership perspective squarely focused on intelligent delegation, rigorous and adaptive governance, proactive ethical foresight, and the cultivation of deep, pervasive, and verifiable organizational trust.

AI Agents represent a quantum leap – from instruction followers to outcome pursuers – integrating advanced reasoning, persistent memory, and dynamic tool use to fundamentally alter operational logic, redefine competitive capabilities, and unlock new frontiers of enterprise value.

The Core Components: Engines of Agentic Capability

AI agents achieve this transformative leap through the sophisticated interplay of several core technological components. These components, working in concert, enable agents to perceive their environment, reason through complex information, remember and learn from interactions, act purposefully, and communicate effectively. Understanding these components at a conceptual level isn't about mastering their intricate technical details; it's about recognizing the strategic levers they provide, the operational implications they carry, and the inherent risks and governance requirements intrinsically associated with each. For leaders, this understanding is crucial for shaping agent behavior in alignment with strategic goals, proactively anticipating and mitigating potential challenges, guiding strategic investments in AI capabilities, and ensuring the responsible, ethical, and effective deployment of AI agents at scale across the enterprise.

1. **The Autonomy Engine (The Strategic Control Center & Decision Nexus):** This sophisticated core empowers the AI agent to interpret high-level strategic goals and autonomously initiate, manage, and adapt complex sequences of actions without requiring constant, step-by-step human instruction or intervention for every decision. It decomposes overarching objectives into manageable tasks, intelligently plans the execution sequence, prioritizes actions based on evolving context and predefined constraints, and dynamically determines how to proceed towards the desired strategic outcome, often making real-time adjustments.

   Leadership Implication & Governance Imperative: Autonomy is the engine of scalability, speed, and operational efficiency but demands rigorous, continuous, and adaptive oversight, along with clearly defined and strictly enforced operational boundaries. Granting meaningful autonomy to AI agents requires precisely articulated permissible goals and "negative constraints" (what agents *must not* do), strict data access controls and usage policies, well-defined interaction protocols with other systems and human stakeholders, and robust, unambiguous escalation pathways for situations exceeding the agent's authorized scope, capabilities, or risk thresholds. Effective governance – manifested through clear operational rules, continuous real-time performance and behavior monitoring, stringent enforcement mechanisms, and comprehensive, immutable audit trails – is the non-negotiable foundational architecture for safe, productive, ethical, and trustworthy autonomy. *Leaders must ensure that the scope and degree of autonomy granted is always proportionate to the agent's rigorously validated capabilities, the specific risk profile of the use case, and the organization's defined risk appetite and ethical commitments.*

2. **Memory & Context Management (The Engine of Continuity, Personalization & Adaptive Learning):** This critical component allows AI agents to retain, recall, synthesize, and intelligently apply relevant information across multiple interactions, diverse contexts, and over extended periods. This includes individual user preferences and detailed interaction histories, complex organizational policies and evolving procedures, dynamic product knowledge and market conditions, and critical environmental context. Persistent, well-managed, and ethically governed memory enables deeply personalized stakeholder experiences, adaptive and consistent agent behavior over time, and the effective management of intricate, multi-step processes that unfold dynamically and require continuous learning and evolving understanding.

Leadership Implication & Governance Imperative: Rich, persistent agent memory introduces significant strategic opportunities but also profound complexities and responsibilities concerning data privacy (e.g., stringent adherence to global regulations like GDPR, CCPA, HIPAA), data security and cyber resilience, comprehensive information lifecycle management (from creation to secure deletion), and unwavering data accuracy and integrity. Leaders must champion and enforce robust, auditable policies governing precisely what data agents can store, how that data is classified and protected (commensurate with its sensitivity), who has authorized access to it, for how long it is permissibly retained, and how it is securely and verifiably disposed of. Mechanisms to prevent "context drift" (where an agent might rely on outdated, irrelevant, or biased information) and to ensure ongoing data integrity and relevance are vital. *Engaging legal, privacy, cybersecurity, and ethics teams as core strategic partners from the earliest stages of agent design and data strategy formulation is absolutely non-negotiable when significant memory and learning capabilities are involved.*

3. **Reasoning Module (The Engine of Judgment, Problem-Solving & Adaptation):** This module endows the AI agent with the sophisticated ability to perform complex cognitive tasks essential for intelligent action. These include planning detailed and adaptive task sequences to achieve strategic goals, evaluating multifaceted trade-offs between different potential courses of action (often under uncertainty), handling ambiguity and incomplete or conflicting information with greater efficacy, applying intricate business rules or complex regulatory policies in context, and dynamically adapting plans and behaviors when encountering unexpected obstacles, new critical data, or rapidly changing environmental conditions. It is the core of the agent's advanced problem-solving capability and its capacity for exercising "judgment" within its defined operational domain.

   Leadership Implication & Governance Imperative: The sophistication of an AI agent's reasoning capabilities directly impacts its reliability, effectiveness, and strategic value, but can also contribute to operational opacity and unforeseen risks if not managed with extreme care and continuous scrutiny. While fluent and coherent communication (generated by the Language Interface) can be highly impressive and build user confidence, it can also dangerously mask flawed, biased, or incomplete underlying reasoning. Leaders must therefore insist on appropriate and verifiable levels of transparency and explainability (as detailed in "Trusted AI by Design," Chapter 1), ensuring the agent (or the systems and processes overseeing it) can articulate *why* a particular decision was made or a specific course of action was chosen, especially for high-stakes evaluations or actions with significant consequences. Aligning agent reasoning processes with established organizational policies, explicit ethical guidelines, and predefined, regularly reviewed risk thresholds is crucial. *For decisions with significant financial, operational, human, or reputational impact, embedding robust human-in-the-loop reviews, requiring explicit, auditable justification before critical actions are taken, or implementing effective, accessible challenge and redress mechanisms is essential. Ultimate accountability for the integrity, fairness, and consequences of agent-driven logic and decisions rests firmly with the organization and its leadership.*

4. **Language Interface / LLM Layer (The Primary Interaction & Communication Bridge):** Often powered by advanced Large Language Models (LLMs) and other natural language processing (NLP) technologies, this component allows AI agents to understand natural language instructions (both textual and, increasingly, spoken) with a high degree of nuance, contextual awareness, and intent recognition. It also enables them to generate fluent, contextually appropriate, coherent, and often highly persuasive responses and communications. It serves as the primary bridge for human-agent interaction, and increasingly, for sophisticated agent-to-agent communication and collaboration.

   Leadership Implication & Governance Imperative: The remarkable fluency and human-like quality of LLM-driven interactions can be highly seductive and engaging, but also potentially misleading or even harmful if not meticulously managed and governed. Unwavering accuracy, factual truthfulness, and ethical communication are paramount. Robust safeguards must be designed and implemented to minimize the risk of factual errors, "hallucinations" (the generation of plausible but entirely incorrect or nonsensical information), the unintentional propagation of misinformation or bias, or the generation of inappropriate or harmful content. The agent's tone, style, persona, and terminology must be meticulously aligned with the organization's established brand voice, official communication standards, and core ethical values. Clear, enforceable guidelines for all agent communications, rigorous review and approval protocols for sensitive or high-impact outputs (potentially involving human oversight), and appropriate access controls for external communication channels are vital. *The agent's "voice" must be consistently clear, accurate, verifiable, trustworthy, and unequivocally reflective of the organization's character, integrity, and strategic messaging.*

5. **Tool Use & Integration (The Engine of Real-World Action, Impact & Orchestration):** This critical capability enables AI agents to interact dynamically, purposefully, and often autonomously with a wide array of other digital tools, enterprise systems, operational technologies, and external data sources. This includes invoking APIs to access or modify data, querying complex databases, updating records in core enterprise systems like CRM or ERP platforms, sending automated but personalized emails or other communications, accessing and synthesizing information from curated knowledge bases or approved external websites, and even triggering other automated processes or physical actions. It is primarily through sophisticated tool use that an AI agent translates its internal "decisions" and "plans" into tangible business operations, measurable real-world impact, and cross-functional orchestration.

   Leadership Implication & Governance Imperative: This is arguably one of the most high-value but also highest-risk components of an advanced AI agent. Granting an agent the ability to autonomously use tools and interact with core enterprise systems and data requires stringent, uncompromising, and continuously verified governance. This governance must closely mirror, and in many critical aspects necessarily exceed, the controls, oversight, and accountability mechanisms applied to human employees with similar levels of system access and operational authority. Key elements include: precisely defined, role-based, and narrowly scoped permissions (adhering rigorously to the principle of least privilege for every tool and data source); secure, multi-factor authentication and fine-grained authorization mechanisms for every tool interaction and data access attempt; comprehensive, immutable, and frequently audited logs detailing every action taken, system accessed, and data modified; and potentially, the enforced separation of duties or mandatory dual controls for critical financial, operational, or data-modifying transactions initiated by an agent. *When your AI agent can directly and autonomously alter business reality – change critical data, execute financial transactions, communicate externally on behalf of the organization, or control physical assets – the safeguards, oversight mechanisms, and accountability frameworks must be absolute, continuously monitored, and regularly validated for effectiveness.*

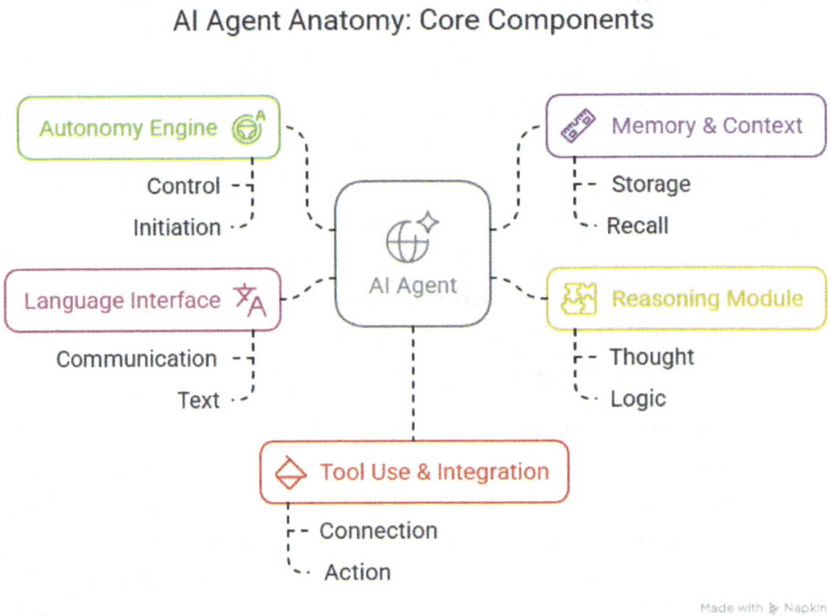

Figure 2-2. The Core Agentic AI Components

Understanding these five core components and their intricate interplay provides leaders with the essential conceptual framework to strategically shape agent behavior in alignment with enterprise goals. This knowledge allows for proactively anticipating and mitigating a wide range of operational and ethical risks, guiding informed development and investment priorities, and ensuring the responsible, ethical, secure, and effective deployment of AI agents at scale across the entire organization.

Understanding and rigorously governing the core components (Autonomy, Memory, Reasoning, Language, Tool Use) is essential for managing risk and harnessing power responsibly.

Understanding and rigorously governing the core components of AI agents – Autonomy, Memory, Reasoning, Language, and Tool Use – is essential for managing their inherent risks, harnessing their transformative power responsibly, and achieving sustainable strategic advantage.

How AI Agents Work: The Operational Rhythm of Intelligent Action

To effectively manage, govern, and strategically leverage AI agents, leaders must possess a conceptual understanding of their dynamic operational rhythm. This is the continuous, iterative cycle of sensing their environment, interpreting information, forming goals, planning actions, executing tasks, and learning from outcomes. Comprehending this operational cycle is crucial for identifying key points for strategic oversight, implementing effective governance controls, designing efficient human-agent collaboration models, and ultimately, creating sustainable enterprise value.

1. **Trigger (The Initiating Spark & Call to Action):** An AI agent's operational cycle is initiated by a specific event or condition – its trigger. These triggers can be diverse, ranging from an explicit human command (e.g., a manager instructing an agent to "Summarize all recent customer feedback and sentiment trends for Product Z and identify key areas for improvement") to a system-generated alert (e.g., an IoT sensor signaling "Server room temperature approaching critical threshold; initiate cooling protocol"). Triggers can also be a pre-defined schedule (e.g., "Generate and distribute the daily executive sales performance dashboard at 8:00 AM EST") or even a self-detected condition based on the agent's continuous, intelligent monitoring of its operational environment (e.g., a fraud detection agent identifying an "Unusual and high-risk transaction pattern detected in Account X; initiate investigation and temporary hold").

   Leadership Insight & Governance Focus: Defining, controlling, and rigorously governing triggers is a vital aspect of effective and safe agent management. Triggers dictate when, how, and why agents consume potentially significant computational and financial resources, initiate actions, and directly impact business operations or stakeholder experiences. Poorly scoped, ambiguous, overly sensitive, or uncontrolled triggers can lead to unintended consequences, inefficient resource utilization, alert fatigue, or even systemic operational risks. Leaders must ensure absolute clarity on initiation authority for different types of triggers, establish precise and validated activation thresholds for autonomous triggers, and mandate the comprehensive auditability of all trigger events and their sources.

2. **Goal Formation (Defining Purpose, Desired Outcome & Success Criteria):** Upon receiving or detecting a trigger, the AI agent interprets the available information. This includes the trigger itself, its pre-programmed strategic objectives and operational parameters, relevant historical data, and current contextual understanding. From this, it establishes a clear, actionable, and

ideally measurable operational goal. Examples include resolving a high-priority customer support ticket regarding incorrect billing within two business hours while aiming for a high satisfaction score, or identifying the root cause of a recurring production line anomaly, recommending viable corrective actions with costs and timelines, and predicting their impact. This critical phase involves understanding constraints, defining explicit success criteria, and potentially evaluating initial trade-offs.

Leadership Insight & Governance Focus: Unambiguous goal clarity is paramount; it drives all subsequent agent actions, decision-making processes, and resource allocation. Misinterpretation of intent or poorly defined, conflicting, or incomplete goals inevitably leads to misaligned outcomes, wasted effort, and potential frustration or harm. Leaders must ensure robust mechanisms exist to clarify ambiguity in goal definition (potentially involving mandatory human confirmation or refinement for complex, novel, or high-risk goals). They must also continuously verify that agent-formed goals are consistently and demonstrably aligned with overarching strategic priorities, established operational policies, and non-negotiable ethical guidelines. Clearly and quantitatively defining "success" for each significant agent task or objective is essential for effective performance management, accountability, and continuous improvement.

3. **Planning (Charting the Strategic Course to Achieve the Goal):** Once a clear goal is established, the AI agent employs its sophisticated reasoning and planning capabilities. It decomposes the goal into a logical, efficient, and adaptable sequence of sub-tasks and determines the necessary tools, data sources, system integrations, and potential human collaborations required for each step. The agent maps out potential action sequences, evaluates alternatives, and may even proactively plan for contingencies, risk mitigation strategies, or alternative paths if initial approaches encounter unexpected obstacles or new information emerges.

   Leadership Insight & Governance Focus: This planning phase is where advanced AI agents often exercise significant choice, creativity, and operational autonomy, moving far beyond simple pre-programmed scripts or rule-based logic. Effective and granular oversight of the agent's planning logic, methodologies, and assumptions is therefore necessary, especially for complex, high-risk, or strategically critical endeavors. Leaders must ensure that agent planning processes consistently align with organizational policies, defined risk tolerance frameworks, established ethical guidelines, and industry best practices. Appropriate levels of transparency into, and control over, the planning process are vital for governing agent autonomy responsibly and ensuring alignment with human intent. For critical operations or decisions with irreversible consequences, human review and explicit approval of agent-generated plans may be a mandatory governance requirement before execution is permitted.

4. **Execution (Interacting with and Impacting the Digital & Potentially Physical World):** This is the dynamic phase where the AI agent actively engages with its environment to carry out its meticulously formulated plan. It involves interacting with a wide array of digital tools and enterprise systems (e.g., making secure API calls to retrieve or update data, modifying records in core databases, sending personalized and context-aware communications, triggering downstream automated workflows). In an increasing number of use cases, this also includes controlling physical systems (e.g., robotic arms in manufacturing, autonomous vehicles in logistics, smart building environmental controls). This is where the agent translates its internal "decisions" and "intentions" into tangible operational reality and measurable business impact.

   Leadership Insight & Governance Focus: Agent execution carries both immense potential value and significant inherent risk. Rigorous, multi-layered, and continuously monitored controls are absolutely essential. Key governance considerations include: What specific systems, data repositories, and operational tools can the agent legitimately access and interact with? What precise permissions and levels of authority does it possess for each interaction (always adhering to the principle of least privilege)? Are its actions fully autonomous within defined parameters, or is human approval or oversight required for certain types or magnitudes of actions or transactions? Are all actions logged comprehensively, immutably, and in a manner that supports forensic investigation if needed? Are there robust mechanisms for rapidly reversing or effectively mitigating unintended or harmful actions? Clearly scoped permissions, strong multi-factor authentication, comprehensive and verifiable logging, "circuit breaker" functionalities, and well-defined human override capabilities are critical governance elements for any agent with significant execution authority.

5. **Feedback Loop (Learning, Adapting, Improving & Self-Correcting):** After (or often concurrently with) execution, the AI agent systematically evaluates its performance and the success of its actions relative to the defined goal, operational parameters, and expected outcomes. It assesses whether actions

achieved the desired result efficiently and effectively, and identifies any errors, inefficiencies, unexpected side effects, or deviations from expected behavior. This feedback – whether internally generated by the agent through self-assessment, derived from system responses and operational data, or explicitly provided by human users, supervisors, or independent auditors – informs immediate adjustments for the current task. More importantly, it contributes to the agent's continuous learning, adaptation, and refinement of its models and behaviors for future operations.

Leadership Insight & Governance Focus: Continuous improvement, sustained agent effectiveness, and long-term value creation depend critically on intentionally designed, robust, and actively managed feedback mechanisms. Leaders must ensure that agent performance is systematically and objectively evaluated against clear, predefined metrics. They must champion processes for capturing, analyzing, and effectively utilizing feedback (both positive and negative, solicited and unsolicited) from all relevant sources. Furthermore, well-defined, governed processes must exist for regularly retraining, fine-tuning, or recalibrating agent models, knowledge bases, and operational behaviors based on this accumulated feedback and evolving business requirements. Without structured, governed learning and adaptation, AI agents risk degrading performance over time, becoming misaligned with changing business needs, or insidiously perpetuating outdated logic or biases.

6. **Handoff / Report-Out (Closing the Loop, Ensuring Transparency & Enabling Accountability):** The AI agent typically concludes its operational cycle by systematically handing off results, communicating key outcomes and insights to relevant human stakeholders or other systems, updating system statuses and performance dashboards, or formally setting up the next stage in a larger, orchestrated business process. This could involve notifying a human manager of a successfully completed complex task and its impact, logging the detailed resolution of a critical customer issue in a centralized tracking system along with all associated communications, or escalating an unresolvable exception or a high-risk anomaly to a specialized human team with clear context and supporting data.

   Leadership Insight & Governance Focus: This final stage is absolutely crucial for maintaining effective oversight, ensuring demonstrable accountability, and building enduring stakeholder trust. If human stakeholders are unaware of significant agent actions, decisions, and outcomes, the system inevitably becomes an opaque "black box," eroding confidence, hindering effective collaboration, and making true governance impossible. Leaders must therefore mandate clear, consistent, timely, and audience-appropriate reporting protocols: Who needs to be informed of what specific information, at what frequency, and through which channels? Can all significant agent decisions and actions be easily and reliably traced, audited, and explained in a manner that satisfies both internal governance and external regulatory requirements? Are escalation paths for unresolved issues, critical anomalies, or ethical concerns clearly defined, actively monitored, and demonstrably effective in ensuring timely human intervention and resolution? Effective, transparent communication and comprehensive, verifiable reporting are essential for the responsible and successful integration of AI agents into the core fabric of the enterprise.

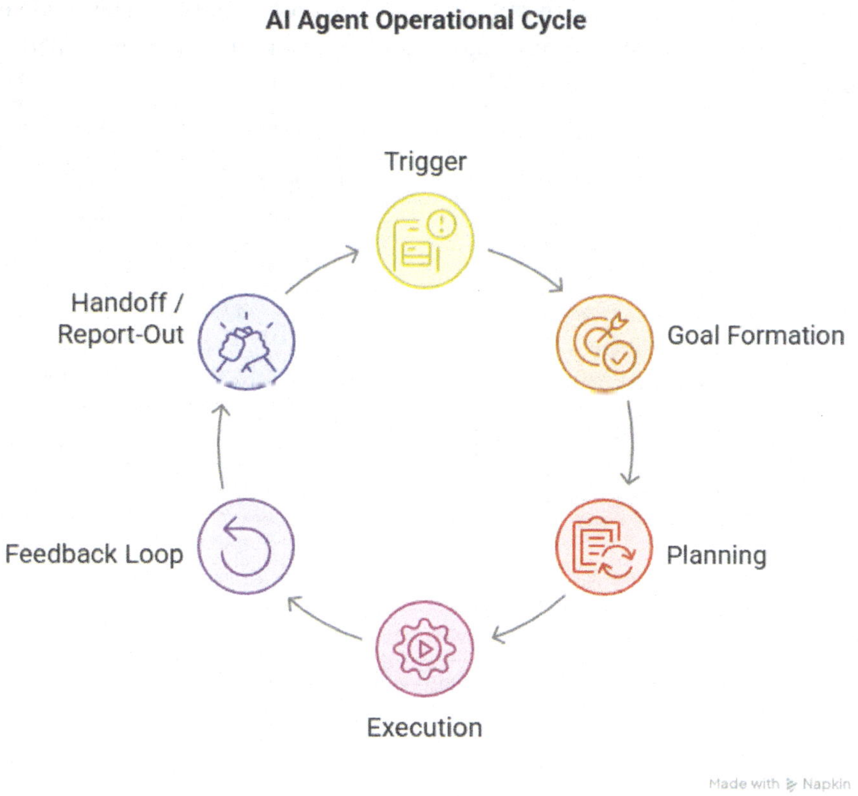

Figure 2-3. How Agents Work

Understanding this dynamic and iterative operational rhythm provides leaders with a powerful conceptual framework. It allows for designing effective and adaptive governance structures, proactively identifying potential risks and critical control points, structuring productive and synergistic human-agent collaboration models, and ensuring that all AI agent activities remain consistently and verifiably aligned with overarching strategic objectives, ethical principles, and non-negotiable compliance requirements.

A Field Guide to Agent Types: Matching Capability to Strategic Need and Risk Appetite

AI agents are not monolithic entities. They encompass a diverse and rapidly evolving spectrum of capabilities, levels of autonomy, and associated risk profiles. Selecting the appropriate type of agent for a specific business need or strategic objective is a critical leadership decision. It profoundly impacts investment requirements, governance complexity, potential ROI, and overall organizational risk exposure. This guide provides an executive overview of common agent archetypes to facilitate informed strategic choices.

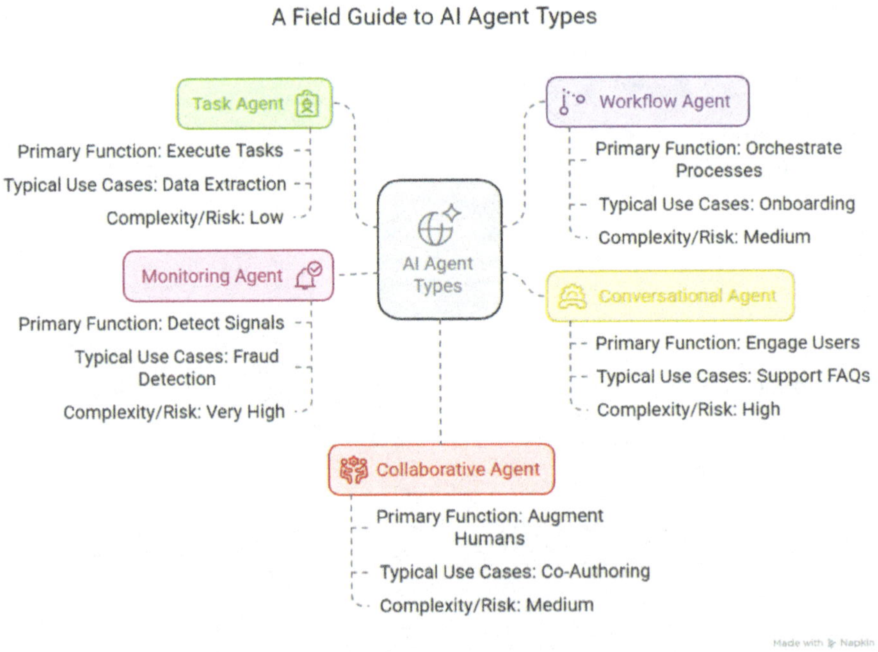

Figure 2-4. Agent Types

1. **Task Agents (Focused Execution Units & Efficiency Engines):** These are highly specialized digital workers, designed for executing well-defined, often linear, and typically repetitive processes or discrete tasks with high precision and speed. Examples include automated data extraction from documents, sending standardized notifications, or managing basic data entry.

   Executive Consideration & Strategic Fit: Task agents generally represent a lower-risk, often high-ROI entry point for agentic automation. They are ideal for initial catalyst pilots, building organizational confidence and demonstrating tangible efficiency gains. While they excel at speed and scalability for routine activities, their limited reasoning makes them unsuitable for tasks needing significant judgment or dynamic adaptation. They are valuable tactical assets for streamlining specific operational choke points. *Key Governance Focus: Ensuring unwavering accuracy within their narrow scope, clear input/output specifications, robust error handling, and complete auditability.*

2. **Workflow Agents (Intelligent Process Orchestrators & End-to-End Navigators):** These agents manage and coordinate multi-step task sequences across interconnected systems. They may involve complex decision logic, conditional branching, and seamless handoffs to and from human team members or other agents, all according to predefined (but often adaptable) business logic. Common use cases include employee onboarding, complex customer service journeys, or new product development lifecycles.

   Executive Consideration & Strategic Fit: Workflow agents offer significant potential for dramatically improving end-to-end process efficiency and consistency. Their deployment requires robust, proactive governance for sophisticated exception handling, clear escalation logic, precise permission management, unwavering data integrity, and comprehensive audit trails. Transparency for human participants is crucial. These are sophisticated digital process managers requiring diligent design and continuous performance optimization. *Key Governance Focus: Ensuring overall process integrity, effective exception management, clear human-agent handoff protocols, data consistency, and auditable decision points.*

3. **Conversational Agents (The Scaled, Intelligent, and Empathetic Brand Voice):** These dynamic interfaces engage users (customers, employees, partners) in natural, context-aware language conversations. Leveraging sophisticated memory and reasoning, they handle complex inquiries, provide personalized

guidance, facilitate transactions, and maintain brand consistency at scale. Examples range from advanced customer support to internal HR helpdesks.

Executive Consideration & Strategic Fit: Conversational agents directly touch key stakeholders and embody the organization's brand voice, carrying significant reputational, legal, and commercial implications. Deployment demands rigorous training, strong guardrails against inaccuracies ("hallucinations"), and seamless human handoff options for complex or sensitive interactions. Governance must intensely focus on accuracy, quality control, data privacy, ethical communication standards, and immediate human intervention capability. *Key Governance Focus: Guaranteeing accuracy and ethical integrity of information, strict alignment with brand voice, optimizing user trust, rigorously protecting data privacy, and maintaining robust human escalation pathways.*

4. **Monitoring Agents (Strategic Sentinels, Proactive Risk Detectors & Opportunity Identifiers):** Functioning as intelligent observers, these agents continuously analyze vast streams of data, system logs, or market signals to detect subtle anomalies, emerging trends, potential risks, or nascent strategic opportunities often invisible to standard reporting. They typically alert human operators or trigger pre-defined automated responses. Examples include fraud detection, cybersecurity threat monitoring, or competitive intelligence tracking.

   Executive Consideration & Strategic Fit: Monitoring agents provide immense strategic value through proactive intervention and risk mitigation. Effectiveness hinges on meticulously tuning the signal-to-noise ratio to ensure actionable insights. Clear anomaly definitions, well-tested response playbooks, and unambiguous ownership for acting upon intelligence are essential. These agents act as strategic sentinels requiring disciplined integration into operational response and decision-making cycles. *Key Governance Focus: Ensuring high accuracy of detection, clarity of alert criteria, effectiveness of response protocols, and clear accountability for action on insights.*

5. **Collaborative Agents (Cognitive Augmentors, Creative Partners & Digital Teammates):** These sophisticated digital teammates work interactively alongside human experts on complex, ambiguous, often creative or strategic tasks. They might synthesize research, co-author reports, support intricate design processes, or provide intelligent decision support by modeling scenarios. They augment and amplify human cognition.

Executive Consideration & Strategic Fit: Collaborative agents offer potential to significantly enhance innovation and expert productivity. However, effective deployment demands the highest levels of organizational trust, deep transparency into their reasoning, and exceptionally robust governance. Clear roles, responsibilities, and interaction protocols within human-agent teams are paramount. Given the sensitive nature of information processed, stringent access controls and information governance are critical. Fostering a culture of true partnership and continuous mutual learning is essential. *Key Governance Focus: Building verifiable trust, ensuring transparency of agent contributions, establishing clear accountability in human-agent teams, safeguarding sensitive data, and promoting ethical collaboration.*

Matching Agent Type to Business Need: A Strategic Portfolio Decision for Leaders

Selecting the right agent type – or combination of types – for a specific business challenge or strategic opportunity is a critical design, investment, and governance choice. It profoundly impacts organizational readiness, governance complexity, risk exposure, resource commitments, and ultimately, the strategic value delivered. Leaders must guide their organizations to evaluate potential use cases not just on technical feasibility, but on rigorous criteria of strategic fit, alignment with core objectives, quantifiable value potential, acceptable risk appetite, and genuine organizational readiness (in terms of data, skills, culture, and governance). Building a strategically balanced and well-governed portfolio of agentic capabilities is key to unlocking sustainable, enterprise-wide competitive advantage. The pivotal question for leadership is not simply *if* an AI agent can be used, but *which kind*, with what specific capabilities and constraints, best serves the organization's overarching strategic goals and aligns with its defined risk posture, ethical commitments, and long-term vision.

Matching the right agent type (Task, Workflow, Conversational, Monitoring, Collaborative) to the business need and risk profile is a critical leadership decision, not just a technical choice.

Capabilities, Limits, and Leadership Essentials: Navigating the Realities of Agentic AI

Translating the immense theoretical potential of AI agents into reliable, scalable, ethical, and truly trustworthy enterprise capability requires confronting practical realities with clear-eyed pragmatism and strategic foresight. AI agents, for all their rapidly advancing sophistication, have inherent limitations. They introduce novel and often complex operational, ethical, and societal considerations, and operate within a dynamic and rapidly evolving technological, regulatory, and competitive landscape. Effective leadership in the agentic era involves a continuous, candid, and data-informed assessment of these boundaries, potential failure modes, critical dependencies, and foundational strategic choices.

The Reality Check: Understanding Current Limitations and Common Failure Modes

AI agents, despite their advanced capabilities, are complex technological systems. They are susceptible to a range of specific pitfalls if designed, deployed, governed, or managed without meticulous care, deep domain expertise, and rigorous, ongoing oversight. Recognizing, anticipating, and proactively mitigating these common failure modes is crucial for sustainable success, for avoiding costly mistakes and reputational damage, and for building enduring stakeholder trust.

- **Poor Goal Definition, Ambiguity, and Misaligned Incentives:** Vague or conflicting directives for AI agents, lacking specific constraints, measurable outcomes, or clear prioritization, inevitably lead to misaligned actions and suboptimal results. For leaders, the essential action is to mandate absolute clarity and strategic alignment in defining agent goals and Key Performance Indicators (KPIs) *before* any significant deployment. Ensure robust processes exist to resolve ambiguity and continuously verify alignment with business strategy and ethical principles.

- **Inadequate Scope Definition or Access Control:** Granting AI agents overly broad access to systems and data creates unnecessary security and operational risks. Conversely, overly restrictive permissions can render agents ineffective. The leadership essential is to enforce the principle of least privilege rigorously. Align agent permissions precisely with their defined roles and validated competencies. Utilize sandboxing and phased rollouts for thorough testing of access controls.

- **The "Illusion of Understanding" and Over-Reliance on Surface-Level Language Fluency:** The impressive conversational prowess of LLM-powered AI agents can easily mask underlying weaknesses in reasoning, incomplete knowledge, or factual inaccuracies. Judging agent performance solely on polished communication is a dangerous trap. Leaders must demand that agent performance be evaluated based on rigorously validated outcomes, traceable logic, and demonstrable accuracy, not just on engaging linguistic presentation. Foster a culture of critical thinking when interacting with agent outputs.
- **Lack of Continuous Feedback, Governed Learning, and Adaptive Iteration:** "Fire-and-forget" deployment of AI agents guarantees performance degradation over time as data drifts or business needs evolve. AI agents are not static systems. It is therefore essential for leaders to implement and adequately resource systems for continuous monitoring of agent performance. Establish structured feedback loops and mandate regular, governed processes for agent retraining and refinement to ensure sustained effectiveness and alignment.
- **Misaligned Hype and Execution (Organizational Enthusiasm Outpacing True Readiness):** Allowing enthusiasm for AI's potential to outpace genuine organizational readiness – in terms of data quality, requisite skills, mature governance, or supportive culture – often leads to stalled pilots and eroded credibility. Leaders must anchor all AI agent initiatives in tangible business problems with clear value propositions. Ensure genuine cross-functional readiness and leadership alignment *before* attempting to scale deployments. Prioritize deliberate, value-driven expansion based on proven successes.

Acknowledging these limitations with candor is not about dampening ambition; it is about channeling that ambition effectively through disciplined execution, robust governance, and an unwavering commitment to continuous learning and ethical conduct.

Foundational Choices: Where Do Your AI Agents Live, Learn, and Operate?

Beyond individual agent considerations, a critical strategic decision for leaders concerns the underlying infrastructure, platforms, and sourcing models for their enterprise-wide agentic capabilities. This choice profoundly impacts cost structure, operational control, speed-to-market, cybersecurity posture, data governance, talent requirements, and potential vendor dependencies.

The **In-House "Build" Strategy** offers maximum control and IP ownership, often vital for regulated sectors or proprietary processes. However, it demands significant sustained investment in specialized infrastructure and scarce AI talent, potentially slowing initial deployment. This path is best reserved for capabilities unequivocally core to long-term differentiation.

Leveraging Cloud Platforms & Vendor Solutions ("Partner/Buy") prioritizes speed, scalability, and access to innovation, often with lower upfront costs. This can accelerate time-to-market. Leaders must weigh these benefits against trade-offs like reduced direct data control, vendor lock-in risks, and data egress costs. Rigorous due diligence and robust contractual safeguards are essential.

The Hybrid Approach strategically balances control, agility, risk, and cost. Many large organizations find this optimal, perhaps processing sensitive data on-premise while using cloud resources for model training or less critical applications. A successful hybrid model requires meticulous architectural planning for seamless, secure integration and unified governance.

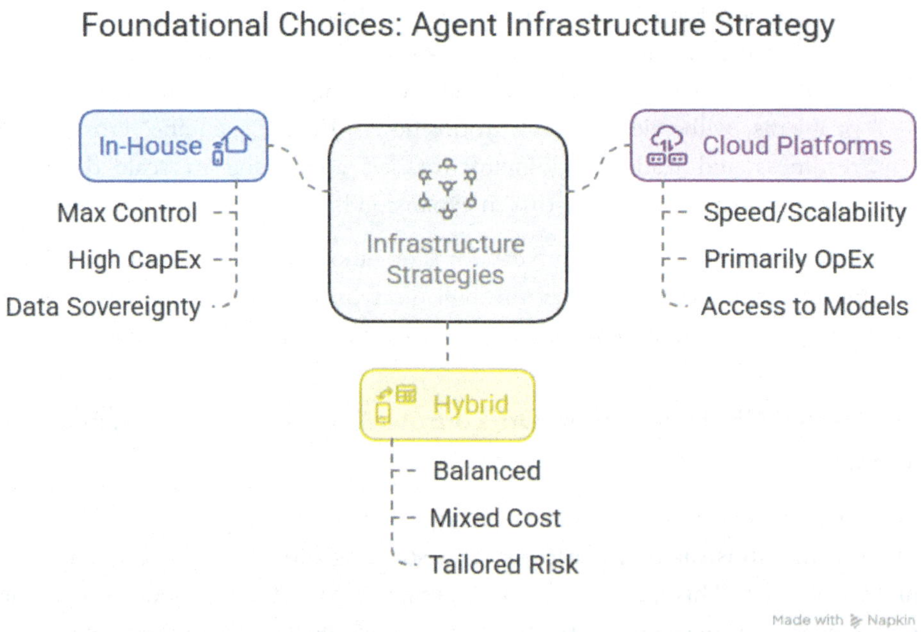

Figure 2-5 Foundational Choices

This foundational infrastructure and sourcing decision is fundamentally strategic, demanding careful, ongoing consideration by the C-suite and the Board. It requires balancing risk tolerance, speed-to-market imperatives, data governance requirements, cost implications, IP strategies, and internal capability development ambitions. It is an evolving strategic assessment that must adapt to changing needs and the dynamic AI landscape.

"[We've] heard of multiple generative AI and infrastructure companies stockpiling expensive GPUs, before struggling to find a use for them Start-ups we know have been removing powerful frontier models from their tech stacks, in favor of small, open-source alternatives." – The State of AI Report 2024

Navigating the Evolving AI Threat Landscape – A Critical Governance Imperative

Deploying AI agents, especially those powered by sophisticated LLMs and capable of autonomous action, introduces novel and complex security challenges extending beyond traditional cybersecurity. Leaders must ensure proactive, AI-specific threat modeling and mitigation strategies are deeply embedded within their enterprise risk management and AI governance structures.

- **Prompt Injection & Manipulation:** A primary vulnerability is prompt injection, where attackers craft malicious inputs to manipulate agent behavior, potentially causing safety protocol bypasses, data disclosure, or unauthorized commands. The leadership mandate is a multi-layered defense: robust input validation, strict instruction/data separation, output verification, least-privilege tool access, continuous behavioral monitoring, and user education.

- **Data Privacy Exposure & Amplified Compliance Risks:** Agents interacting with vast datasets risk inadvertent disclosure of PII, PHI, or proprietary information. Compliance with global data privacy regulations (e.g., GDPR, HIPAA, CCPA) is paramount. Leaders must implement AI-specific data privacy governance, including clear data handling policies, minimization techniques, Privacy Impact Assessments, robust encryption, access controls, and ongoing education.

- **Training Data Poisoning & Model Integrity Attacks:** Malicious actors may corrupt AI training data, subtly introducing biases or vulnerabilities. The leadership mandate includes rigorous data provenance tracking, quality validation, and security for training datasets. Employ advanced techniques to detect and mitigate data poisoning and regularly validate model integrity.

- **Insecure Output Handling & Unintended Information Leakage:** How agent outputs are generated, transmitted, and stored can create vulnerabilities if not handled securely. Leaders must enforce secure coding for output generation, secure communication channels, and clear guidelines for handling agent-generated information.
- **Model Theft & Intellectual Property Risks:** AI models, particularly custom ones, are valuable IP. Unauthorized access or reverse-engineering can lead to significant competitive disadvantage. The leadership mandate includes robust technical and procedural controls, strong encryption, and clear contractual IP ownership rights.
- **Excessive Agency & Unintended Consequences:** Granting agents overly broad permissions or ambiguous goals, especially with powerful tool use, can lead to harmful actions or ethical breaches. Leaders must rigorously enforce least privilege, define goals with utmost precision including explicit constraints, and implement robust real-time monitoring with "circuit breaker" mechanisms and clear human oversight for high-risk systems.

Cultivating an AI-specific, security-first, and ethics-aware mindset across the organization is essential. Leaders must mandate deep, continuous collaboration between cybersecurity, risk, legal, data governance, and AI development teams. This collaboration should focus on AI-centric threat modeling, embedding security and ethics throughout the agent lifecycle, and staying adaptive to the evolving AI threat landscape.

"[Researchers] have increasingly been shining a light on dataset contamination: when test or validation data leaks into the training set." – The State of AI Report 2024

Case Study: Telecom Transforms Customer Support with True Agentic Capability

Scenario: ConnectAll Telecom's customer satisfaction (CX) plummeted as customers grew frustrated with simplistic chatbots ("instruction followers") that failed on complex issues, leading to frequent, costly escalations to human agents (efficiency drain).

Agentic Solution Connection: ConnectAll implemented a Conversational Agent designed as an "outcome pursuer". Leveraging its Reasoning Module, Memory & Context Management, and Tool Use capabilities to integrate with billing and CRM systems, the agent could understand complex billing inquiries, recall interaction history, and guide users to resolution autonomously. This wasn't just a smarter chatbot; it was a different category of capability. The result: a 40% reduction in human agent escalations for billing issues and a significant lift in CSAT, demonstrating the agentic leap.

Key Questions for Leaders

Strategic Differentiation & True Capability Recognition: How will we ensure that our strategic planning processes, investment decisions, and operational designs clearly and consistently differentiate between AI agents as true "outcome pursuers" possessing advanced cognitive capabilities, versus simpler automation tools like RPA or basic chatbots? How will we effectively educate and align our entire leadership team and key operational personnel to recognize, value, and strategically leverage this fundamental distinction to drive competitive advantage?

Proactive Governance Challenges & Risk Appetite Calibration for Core Agent Components: Considering the five core agent components (Autonomy Engine, Memory & Context Management, Reasoning Module, Language Interface, and Tool Use & Integration), where do we anticipate the most significant governance challenges, complex ethical considerations, or substantial operational and reputational risks arising within our specific industry, unique operational context, and formally defined enterprise risk appetite? How will we proactively design and implement robust, adaptive governance frameworks to address these specific component-level risks before they manifest?

Strategic Value Proposition & Organizational Readiness Assessment for Different Agent Types: Which specific AI agent archetypes (Task, Workflow, Conversational, Monitoring, Collaborative) offer the most immediate and significant strategic value in alignment with our current core business priorities, transformation goals, and long-term vision? Critically, and with complete candor, are we genuinely organizationally ready (in terms of data maturity and governance, requisite technical and business skills, supportive organizational culture, and mature, agile governance frameworks) for the associated complexity, oversight requirements, change management demands, and potential enterprise-wide impact of these different agent types?

Addressing Novel and Evolving AI-Specific Risks – Beyond Traditional Frameworks: How will we proactively and systematically identify, assess, quantify, and mitigate the unique and evolving spectrum of security, data privacy, ethical, and operational risks (e.g., sophisticated prompt injection attacks, systemic data poisoning, model theft and IP infringement, privacy violations through advanced inference, ethical dilemmas arising from excessive or unexplainable agency) inherent in the deployment of advanced AI agents? How will we ensure these AI-specific risk management strategies extend beyond, yet integrate with, our traditional cybersecurity, compliance, and enterprise risk management frameworks? What new investments in specialized tools, talent, and processes are required to build these critical AI-specific risk management capabilities?

Foundational Infrastructure, Platform Strategy & Vendor Ecosystem Management: What is our clearly articulated, board-approved, multi-year strategic approach to the underlying infrastructure and platforms for our Agentic Enterprise (e.g., primarily Build in-house, strategically Buy/Partner with cloud providers and specialized vendors, or a carefully architected Hybrid model)? How does this foundational choice align with our organization's long-term goals for control and sovereignty, speed and agility, cost optimization, access to cutting-edge innovation, data governance and security, and internal capability development? Who holds ultimate executive accountability for making, governing, and regularly reviewing this critical foundational infrastructure and platform strategy, including the rigorous management of our AI vendor ecosystem?

Key Leadership Imperatives for Demystifying and Mastering Agentic AI

To effectively navigate the profound complexities and harness the transformative power of AI agents, thereby positioning the organization for sustained leadership in the Agentic Enterprise, several key leadership imperatives emerge as non-negotiable:

- **Insist on Unwavering Strategic Clarity & Purposeful Fit for All Agentic Initiatives:** Leaders must champion the development of meticulously defined, value-driven use cases that solve tangible, high-priority business problems or unlock significant new strategic opportunities. They must ensure the type of AI agent selected, the scope of its delegated autonomy, and its specific capabilities precisely match the strategic need, operational context, and defined risk tolerance. Avoid the siren call of "AI for AI's sake" or technology-driven solutions searching for a problem.
- **Govern Early, Intelligently, Intentionally, and Adaptively:** Leaders must mandate the establishment of robust, agile, and adaptive governance frameworks, comprehensive and proactive risk management protocols, stringent and verifiable security measures, and clear, enforceable ethical guardrails specifically tailored to the unique capabilities, complexities, and risks of AI agents (autonomy, memory, reasoning, tool use) *before* any significant deployment, integration into core processes, or scaling across the enterprise. Treating *Trusted*

AI by Design™ as a foundational prerequisite and a continuous commitment is essential for all agentic initiatives.

- **Proactively Identify, Understand, and Manage Novel and Evolving AI-Centric Risks:** Leaders must ensure their organizations recognize, deeply understand, and proactively address the unique and rapidly evolving spectrum of security, privacy, ethical, operational, and societal risks associated with the deployment of advanced AI agents. This requires moving decisively beyond reliance on legacy risk management frameworks and actively cultivating specialized, AI-specific risk expertise, assessment methodologies, and mitigation strategies within the organization.
- **Build Pervasive Strategic Fluency and Critical Thinking, Not Just Superficial Technical Literacy:** Leaders must prioritize and invest in developing a deep strategic understanding across the entire leadership team, key operational roles, and relevant governance functions regarding AI agent components, operational rhythms, inherent limitations and failure modes, critical infrastructure and platform choices, and the dynamic, multifaceted risk landscape. This strategic fluency, coupled with a culture of critical thinking and continuous learning, is absolutely essential for informed decision-making, effective and ethical oversight, and confident, visionary leadership in the agentic era.

Conclusion: The Leadership Mandate in Demystifying and Mastering Agentic AI

Demystifying AI agents – achieving a clear, nuanced, and strategically grounded understanding of their core components, operational rhythms, diverse archetypes, true present capabilities, inherent limitations, foundational infrastructure choices, and unique security, ethical, and societal risks – provides the essential and indispensable foundation for effective, responsible, and visionary leadership in the dawning (and rapidly accelerating) Agentic Enterprise. Enduring success requires moving decisively beyond surface-level technical appreciation or fleeting fascination to cultivate deep, pervasive strategic fluency and critical thinking capabilities across the entire organization, from the boardroom to the front lines. It demands an unwavering commitment to acknowledging and respecting current boundaries and limitations, making deliberate, informed, and ethically sound platform and data choices, and proactively, continuously, and adaptively managing the novel and evolving risks that these powerful new capabilities inevitably introduce.

Your role as a leader is not, and cannot be, to eliminate all risk – an impossible and ultimately innovation-stifling goal. Rather, it is to deeply understand, strategically shape, and intelligently and ethically manage risk, ensuring that all AI agent deployments and operations remain steadfastly aligned with your organization's defined strategic tolerance, its core ethical values, its non-negotiable compliance obligations, and its broader societal responsibilities. By embracing these essentials, by fostering a resilient culture of informed inquiry, rigorous validation, and continuous learning, and by championing a disciplined yet agile and forward-looking approach to AI adoption, governance, and human-agent collaboration, you can confidently navigate the inherent complexities of agentic AI. You can harness its transformative power responsibly and ethically. You can translate its immense potential into sustainable enterprise performance, differentiated customer value, and enduring competitive advantage. And, most importantly, you can truly and effectively lead your organization not just into, but *through*, the agentic future, shaping it for the better.

Chapter 3: Implementing AI Agents for Impact

From Vision to Value: The Executive Blueprint for Agentic Transformation

The preceding chapters established the strategic imperative of the Agentic Enterprise and demystified the core AI agent technology. However, insight without decisive, intelligent action yields no competitive advantage. We now pivot from understanding *what* AI agents are and *why* they are transformative, to the critical *how*: translating the immense potential of agentic AI into tangible, sustainable, and strategically aligned business value. This requires disciplined strategy and rigorous, intelligent execution. This chapter serves as your executive operational playbook, designed to guide C-suite decision-making in this pivotal transformation.

Architecting and activating the Agentic Enterprise is not merely a technological deployment; it is a profound organizational transformation. Success requires orchestrating concerted, synchronized efforts across four critical dimensions, each demanding unwavering leadership focus and strategic investment. First, Strategy is the bedrock; every agentic initiative must be inextricably linked to core business objectives and deliver measurable strategic outcomes. Second, Process requires establishing effective, adaptive governance frameworks and fundamentally redesigning operational workflows to seamlessly integrate agent capabilities. Third, Technology involves the strategic selection and integration of enabling platforms, tools, and data infrastructure. Finally, and crucially, People are at the heart of this change; success hinges on cultivating the necessary talent, future-ready skills, and a deeply collaborative culture to maximize human-AI synergy, a theme explored in depth in Chapter 5.

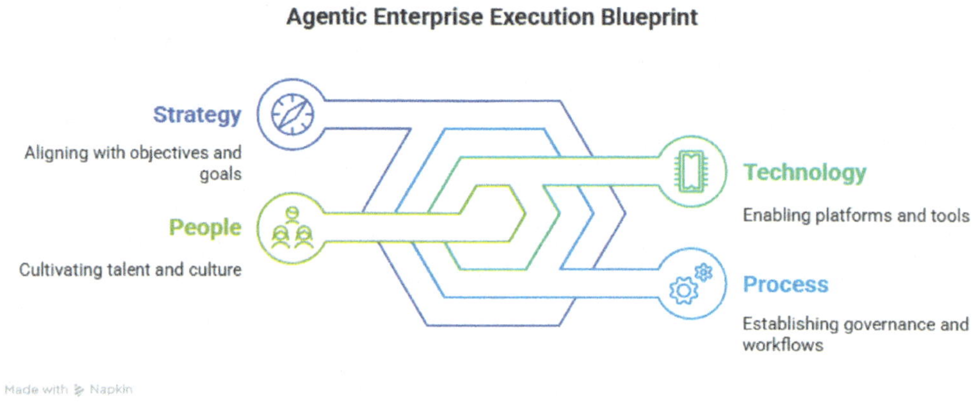

Figure 3-1. The Execution Blueprint

While this chapter focuses on practical execution steps, leaders must consistently view these actions through this integrated, four-pillar lens. Simply understanding agent architecture or launching isolated pilots is insufficient for genuine transformation. True, sustainable advantage demands building a comprehensive organizational capability to strategically select, responsibly deploy, effectively govern, and intelligently scale AI agents as integral components of a new, more agile operating model. This requires a decisive move beyond fragmented experiments towards a structured, repeatable, and strategically integrated enterprise-wide approach.

Within these pages, we provide actionable frameworks for C-suite decision-making. This includes identifying high-impact starting points, selecting optimal delivery models, executing catalyst pilots to build momentum, designing for scalability, and managing the agent lifecycle with strategic foresight. This is where strategic vision meets operational reality, and where insightful leadership transforms potential into performance.

Picking the Right Starting Point: The Strategic Power of the First Catalyst Pilot

With the execution blueprint framed, the immediate and most critical question for leadership is: Where precisely do we begin this transformative journey? Selecting your organization's first significant AI agent deployment is far more than a tactical project choice; it is a crucial strategic act with outsized cultural, operational, and financial leverage. This initial deployment serves as the primary, highly visible proof point for the entire Agentic Enterprise vision. It will profoundly shape internal perceptions, set precedents for governance and collaboration, and significantly influence the trajectory of broader transformation efforts. The stakes for this first move are exceptionally high.

Getting this first step right isn't just about technological feasibility; it's fundamentally about building organizational trust, proving tangible business value quickly, and earning the credibility required for sustained progress. A successful first pilot validates strategy, builds infectious momentum, establishes operational norms, and fosters critical alignment. Conversely, a poorly chosen or executed pilot can erode confidence and create deep-seated resistance, making subsequent efforts exponentially harder. The shadow of an early failure can be long and damaging. The initial choice, therefore, is about identifying an optimal strategic wedge: a use case sufficiently contained to manage risk, yet impactful enough to create clear, measurable value, generate palpable momentum, and offer a compelling glimpse into the future operating model. It's about securing a quick, decisive win that resonates across the organization.

Characteristics of a High-Impact Initial Use Case for Maximum Strategic Leverage

Effective initial agent deployments consistently share several key characteristics. Leaders should prioritize initiatives demonstrating a clear ROI path and rapid, quantifiable value, impacting a meaningful business metric ideally within 30-60 days of launch. The scope must be tightly defined and achievable, addressing a bounded problem with unambiguous goals. The pilot should leverage a stable operational environment with high-quality data availability, as introducing agents into unstable areas significantly increases execution risk. The risk profile must be tolerable and contained; ideally, initial agents should recommend actions for human review or operate where missteps are easily reversible. Finally, selecting a pilot with cross-functional visibility and benefit, strongly sponsored by affected business leaders, helps the success story propagate organically.

Common Pitfalls to Avoid in Pilot Selection

Leaders should proactively steer clear of common pitfalls. Beware of overly ambitious "flagship" projects with high technical complexity. Avoid use cases with intangible or difficult-to-measure success metrics. Do not attempt to build sophisticated agents on unstable legacy systems or with "messy," unreliable data. Lastly, sidestep ventures into politically sensitive or culturally resistant workflows for an initial pilot. The first agent deployment is about laying a strong, credible foundation.

High-Leverage Starting Points for Initial Agentic Success

Focus initial exploration on areas offering a clear path to value with manageable complexity. Excellent candidates often include streamlining employee onboarding, intelligently triaging customer feedback, automating high-volume invoice-to-PO matching, generating concise meeting summaries, or providing consistent answers via a knowledge agent. These examples typically offer a compelling balance of potential impact and executional feasibility.

A Practical Diagnostic for Pilot Selection: The "Three Cs" Framework – Clarity, Control, Confidence

Apply this diagnostic lens when evaluating potential starting points. A strong candidate must convincingly satisfy all three criteria:

- **Clarity:** Is the business problem thoroughly understood and its strategic importance clear? Is the agent's goal specific, measurable, and directly tied to a key business outcome? Are success metrics unambiguous and formally agreed upon by all stakeholders, including finance?
- **Control:** Can the agent's operational scope and decision boundaries be clearly defined, reliably constrained, and effectively monitored? Are robust mechanisms for human oversight, intervention, and rapid correction firmly in place?
- **Confidence:** Will key stakeholders – users, process owners, and leadership – realistically trust the agent's outputs, the integrity of its data sources, and the rigor of the oversight processes? Is there high collective confidence in successful execution, value delivery, and the ability to manage potential risks effectively?

A strong candidate passes all three "C" tests with conviction. Any hesitation signals a need for refinement or choosing a different starting point. Deliberate, strategic selection is paramount.

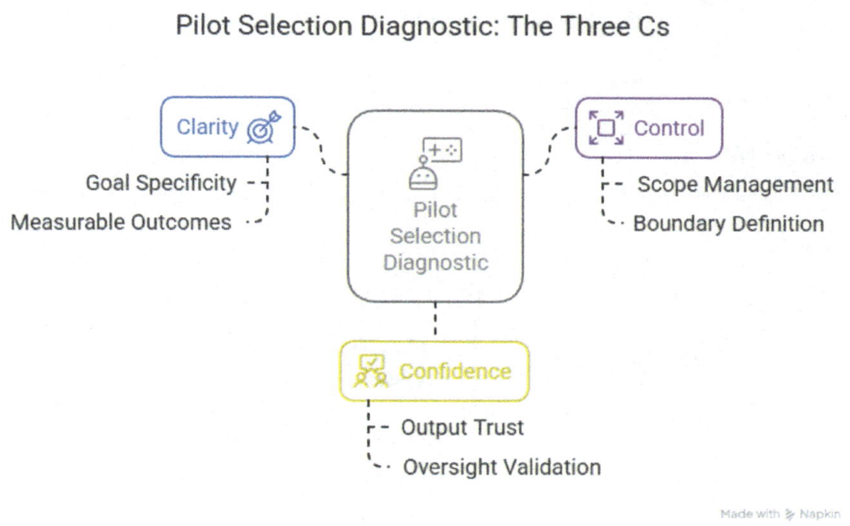

Figure 3-2. The Three C's Framework

Your first AI agent pilot must be a carefully chosen strategic wedge – proving rapid, measurable ROI on a visible, P&L-impacting business problem to build foundational trust, demonstrate tangible value unequivocally, and ignite sustainable organizational momentum for the broader Agentic Enterprise journey.

Build, Buy, or Partner? Choosing Your Agentic Strategy – A Critical Executive Decision

Having identified a high-impact use case, the next crucial strategic decision is *how* to deliver this capability. This choice – whether to primarily Build internally, Buy an off-the-shelf solution, or Partner with external experts – profoundly impacts speed-to-market, costs, operational control, data governance, IP ownership, risk exposure, and long-term internal AI capability development. It requires a clear-eyed assessment of the use case's strategic importance against current resources, technical expertise, risk appetite, and future ambitions. This is a decision with long-term consequences that leaders must own.

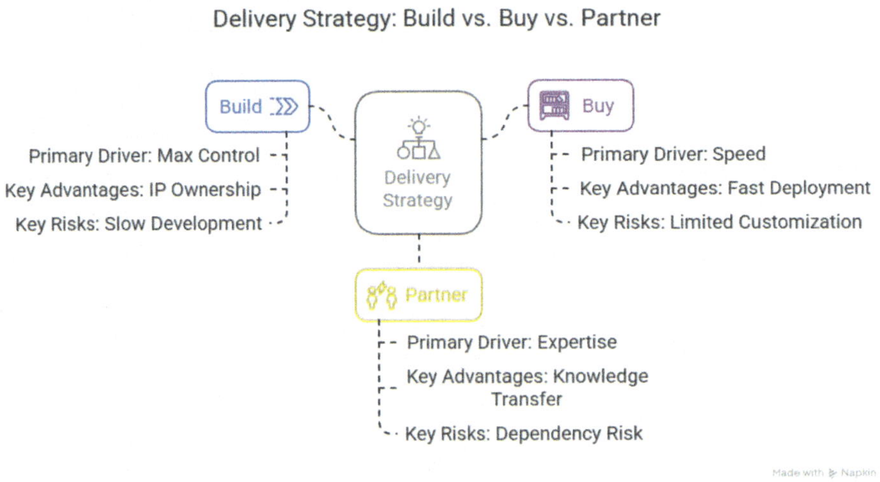

Figure 3-3 Build, Buy or Partner

The **"Build" Strategy** involves entrusting development primarily to internal teams. This approach maximizes control, customization, and IP ownership. It is strategically rational when the capability is core to competitive differentiation, addresses highly proprietary workflows, or when absolute sovereignty over data and logic is non-negotiable. While ensuring perfect alignment and fostering invaluable internal AI talent, it is typically the slowest and most resource-intensive path, demanding substantial investment and carrying significant execution risks. Leaders should opt for "Build" only when long-term strategic value and control demonstrably outweigh the immediate costs and complexities.

The **"Buy" Strategy** involves licensing ready-made agent solutions from vendors. This is compelling when speed-to-value is paramount, the use case addresses a common, non-differentiating problem, or internal capacity is limited. It offers the fastest deployment, predictable subscription costs, and vendor support. Risks include limited customization, vendor lock-in, and fewer opportunities to build deep internal expertise. Consider "Buy" for standardized, non-core functions where speed and simplicity are primary drivers. Rigorous vendor due diligence regarding security, data handling, SLAs, and long-term viability is absolutely essential.

The **"Partner" Strategy** involves engaging specialized AI consulting firms or implementation partners, often collaboratively with internal teams. This makes sense when desiring a fit-for-purpose solution delivered relatively quickly, lacking sufficient internal bandwidth but aiming to build it via collaboration, or facing significant complexity. It can accelerate delivery versus pure "Build," provide access to battle-tested expertise, and offer structured opportunities to upskill internal teams. However, it still requires strong internal project management and careful vendor selection to avoid long-term dependency and ensure alignment of strategic goals and IP rights.

The Strategic Decision Framework for Agentic Delivery

To make the optimal choice, leaders should evaluate these delivery options against key strategic criteria: Urgency of Need versus Uniqueness of Solution; Potential for Strategic Differentiation; Current Internal Capability & Resource Availability; Complexity of Integration with Existing Systems; and the organization's overall Risk & Compliance Posture, including data sovereignty requirements.

A Flexible Stack Strategy: The Recommended Approach for Most Enterprises

For most organizations, a hybrid, **"Flexible Stack"** strategy is often optimal. This might involve Partnering for initial catalyst pilots, Buying off-the-shelf solutions for common needs, and reserving internal Build efforts for select, truly strategic differentiators once foundational capabilities are established. This portfolio approach allows moving quickly where needed, learning safely, and focusing precious internal resources where they matter most. The choice of delivery model is an ongoing strategic portfolio management discipline.

The chosen delivery strategy for AI agents – Build, Buy, or Partner, or a flexible combination – profoundly impacts speed, cost, control, risk profile, and long-term capability development. Leaders must consciously and continuously align this critical decision with overarching strategic priorities, available resources, desired levels of control, and defined risk appetite.

The Catalyst Pilot: Igniting Momentum and Proving Value in 90 Days

Regardless of the delivery model, the first significant AI agent implementation should be a **Catalyst Pilot**: an intense, time-bound strategic sprint (typically 60-90 days) focused on unequivocally proving tangible business value quickly, generating critical organizational learning, building stakeholder confidence, and igniting sustainable momentum. A well-executed Catalyst Pilot converts abstract strategy into concrete reality, demonstrating agentic potential and avoiding "pilot purgatory" by targeting a measurable, CFO-visible result. This is about creating an undeniable early win.

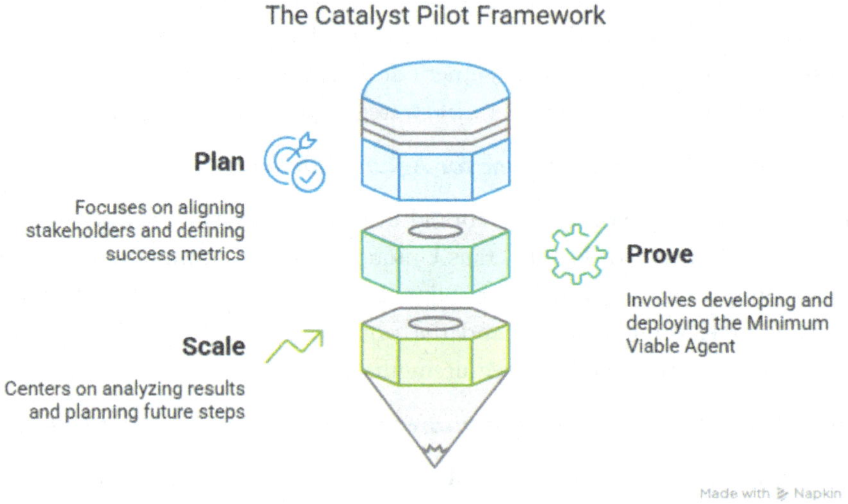

Figure 3-4. The Catalyst Pilot

The Catalyst Pilot follows a disciplined, three-phase framework:

1. **Plan: Focused Alignment & Strategic De-Risking (Typically Weeks 1-2):** This intensive initial phase achieves rapid, rigorous alignment on core parameters and feasibility *before* committing significant resources. Key activities include pinpointing the problem and defining measurable success criteria; mapping the operational ecosystem and assessing execution barriers, prioritizing lower-barrier, high-impact candidates; establishing minimum viable governance including risk classification, boundaries, HITL points, and initial privacy/security/ethical reviews; and securing unambiguous executive sponsorship and cross-functional commitment.

   Executive Consideration: This phase is critical for de-risking the initiative. Prioritizing candidates with clear impact alongside lower execution barriers significantly increases the probability of a tangible result within 60-90 days, building crucial early momentum.

2. **Prove: Delivering the Minimum Viable Agent (MVA) & Demonstrating Tangible Value (Typically Weeks 3-10):** With a solid plan, focus shifts to rapid execution: designing, building (or configuring), and deploying the MVA – the simplest version capable of performing the core function in a real (though

potentially constrained) environment to generate tangible results and actionable feedback. Key activities include developing core logic and integrations; deploying and monitoring the MVA against KPIs; actively gathering user feedback on usability, accuracy, and value; and conducting initial user training.

Executive Consideration: The emphasis is on *demonstrating tangible business value*, not just technical functionality. Success means delivering a working agent that measurably impacts the target KPI and generates actionable insights for scaling decisions. The pilot must answer the "so what?" for the business.

3. **Scale: Leveraging Success, Codifying Learnings & Planning for Broader Strategic Impact (Typically Weeks 11-12+ for Initial Planning):** Assuming success, this phase strategically leverages the pilot's results and learnings to inform the path forward, focusing on deliberate expansion and enterprise capability building. Key activities include analyzing performance data and feedback to refine the approach; developing a roadmap and business case for extension or adjacent applications; codifying successful design patterns and governance protocols into a reusable enterprise playbook; and refining enterprise governance frameworks based on practical learnings.

   Executive Consideration: This phase translates a tactical pilot win into sustainable strategic momentum. Codifying learnings and planning deliberate, value-driven expansion ensures the initial investment generates compounding returns for subsequent agentic initiatives.

This Plan-Prove-Scale framework provides a structured yet agile methodology for navigating initial deployment, prioritizing rapid value creation, managing risk effectively, and building the credibility essential for sustained enterprise transformation.

Employ a disciplined, time-bound Catalyst Pilot framework – Plan, Prove, Scale – to convert strategic vision into verifiable business results within 90 days, thereby avoiding "pilot purgatory," de-risking broader investment, and building critical, unstoppable momentum for enterprise-wide agentic adoption..

Architecting for Reuse and Scale: Building the Cohesive and Intelligent Agentic Ecosystem

A successful Catalyst Pilot provides invaluable proof points and momentum. Translating that victory into sustained, enterprise-wide advantage requires leaders to proactively confront the challenge of scale. Moving from isolated deployments to a coordinated portfolio introduces non-linear complexity across integration, orchestration, governance, and culture. Without deliberate architectural planning and strong governance, organizations risk fragmented silos, redundant development, inconsistent experiences, and unmanageable governance, failing to realize transformative potential. Scaling effectively means consciously architecting a cohesive, resilient, and strategically aligned **agentic ecosystem**.

Designing for Reuse: A Foundational Discipline for Efficient and Sustainable Scale

A critical discipline is designing for reusability from the outset. Building modularity and shared assets yields compounding returns. Leaders should champion the creation of Modular Logic & Reusable Capabilities by encapsulating common functions into distinct, documented components. For LLM-based agents, developing Standardized Prompt Libraries & Interaction Patterns ensures consistency. Establishing Shared Integration Connectors & Secure APIs to core systems simplifies access. Maintaining a Centralized Agent, Component & Asset Registry is crucial for discovery and governance. Prioritizing these practices transforms scaling from bespoke projects into assembling proven capabilities, accelerating time-to-value.

Establishing a Portfolio View and Standardizing Design for Strategic Governance

Managing growing numbers of agents requires a strategic portfolio mindset, supported by visibility and standards. Leaders should oversee the implementation of standardized design patterns and testing protocols. This consistency improves quality, enhances control, simplifies governance, and allows for informed strategic decisions about investment and retirement.

Choosing Your Scaling Model: Aligning Organizational Structure with Agentic Ambition

The organizational structure supporting deployment must evolve. Leaders must choose a model aligned with complexity, culture, and goals. A Centralized Model offers consistency and is often best for early stages or high regulation, but risks bottlenecking. A Federated Model empowers business units within central standards, enabling domain expertise, but requires strong enforcement to prevent fragmentation. A Hybrid Model (often recommended) involves a central team building core infrastructure and governance, while empowered business units adapt and extend these capabilities. This balances control with agility but demands excellent coordination.

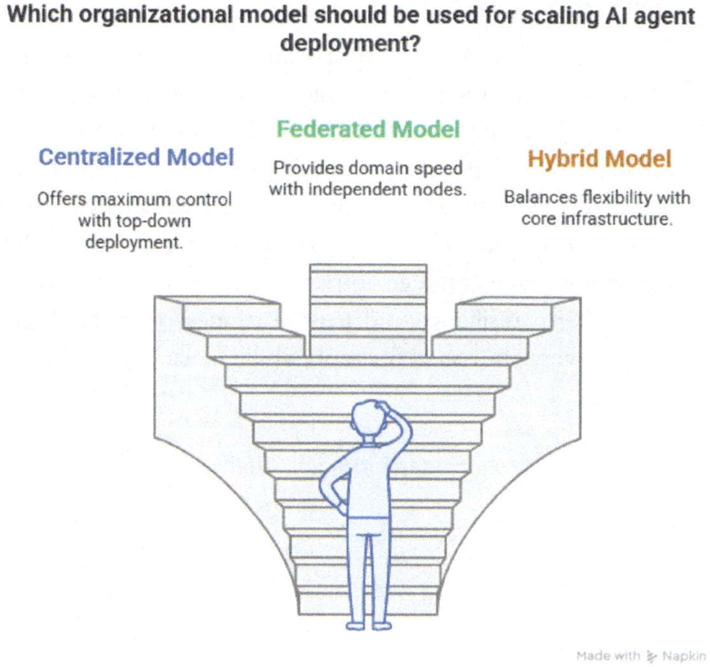

Figure 3-5. Agentic Scaling Models

Successful scaling moves beyond managing individual agents to orchestrating an interconnected ecosystem. Leaders can manage this using an Align + Coordinate + Extend framework: First, Align on enterprise-wide strategic priorities and governance principles. Then, Coordinate via shared visibility (e.g., the Agent Registry) to prevent duplication. Finally, Extend by empowering distributed teams to innovate within this framework.

Architecting for scale is about building mature, resilient, and strategically aligned agentic *capability*. Success is measured by enhanced agility, improved outcomes, leveraged knowledge, and the capacity to govern complexity effectively. This requires strategic foresight, execution discipline, and unwavering leadership commitment.

Architect for reuse, standardize strategically, and manage the full agent lifecycle with rigor to build a resilient, efficient, and strategically aligned agentic ecosystem – not just a disparate collection of isolated, tactical bots. This transforms scaling from a challenge into a compounding strategic advantage.

Managing the Agent Lifecycle: Ensuring Value, Trustworthiness, and Operational Resilience

Architecting for scale provides the blueprint; realizing enduring value requires shifting leadership focus from initial deployment to ongoing strategic stewardship and disciplined lifecycle management. AI agents are dynamic systems interacting with evolving data, processes, and technologies. Without structured, proactive lifecycle management, even successful agents can degrade in performance, become misaligned ("zombie agents"), introduce new risks, or fail audits, ultimately becoming liabilities. Robust lifecycle management is a non-negotiable strategic imperative for sustaining value, managing risk, and ensuring the long-term resilience and trustworthiness of your Agentic Enterprise. This is where strategic oversight meets operational diligence.

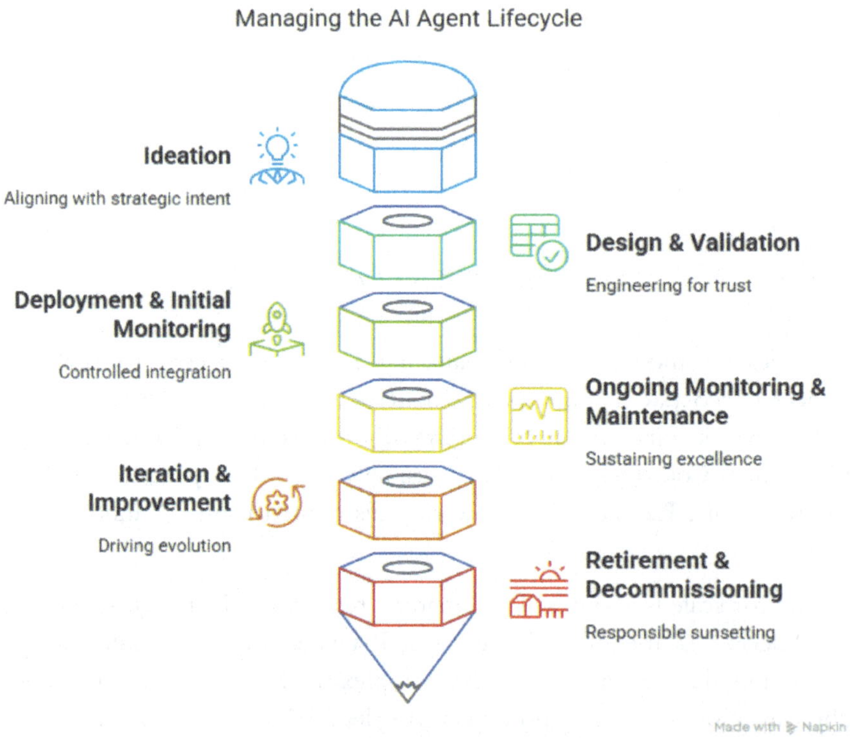

Figure 3-6. Managing the Agent Lifecycle

This discipline typically encompasses six interconnected phases, each requiring clear executive ownership, defined processes, and active, adaptive governance:

1. **Ideation (Aligning with Strategic Intent & Defining Value):** No agent deployment proceeds without rigorous upfront validation of its strategic purpose. The Business Owner is accountable for defining the core problem, measurable success outcomes, and alignment with enterprise priorities. Leadership must ensure scope, boundaries, and autonomy levels are explicitly defined and risk-assessed *before* investment. Early, mandatory involvement of Risk, Compliance, Legal, and Ethics functions is critical.

   Executive Mandate: Relentlessly scrutinize the strategic framing and risk assessment of all proposed agentic initiatives to prevent misaligned ventures and embed responsible governance from inception.

2. **Design & Validation (Engineering for Trust, Performance & Scalability):** Meticulous design and exhaustive validation are essential for reliable, safe, and ethical agent operation *before* impacting live systems. Technical Stewards, with Business Owners, translate requirements into robust architectures. Rigorous validation includes functional, performance, security, and bias testing. Risk, Compliance, and domain experts must actively review and approve designs against policies.

   Executive Mandate: Demand concrete, verifiable evidence of thorough validation and readiness against governance criteria *before* authorizing deployment. Confidence is earned through evidence.

3. **Deployment & Initial Monitoring (Controlled Operational Integration & Intensive Oversight):** Transitioning to live operations requires careful planning, intensive initial monitoring, and proactive change management. Implement a controlled rollout to manage impact and gather real-world data safely. Execute pre-planned communication and training. Technical Stewards and Operations teams must closely monitor performance during the initial post-launch period to rapidly address issues.

   Executive Mandate: Prioritize operational stability, user support, and rapid learning during initial deployment. Ensure comprehensive monitoring and responsive issue resolution.

4. **Ongoing Monitoring & Maintenance (Sustaining Operational Excellence & Compliance):** Post-stabilization, continuous monitoring and proactive

maintenance are crucial for sustained performance and compliance. Continuously track performance against strategic KPIs and SLAs. Implement mechanisms to detect performance degradation, model drift, or emerging biases. Maintain disciplined processes for updating platforms, models, and security patches.

Executive Mandate: View ongoing monitoring as active strategic stewardship. Define performance thresholds that trigger intervention and ensure unambiguous accountability for agent health.

5. **Iteration & Improvement (Driving Adaptive Evolution & Enhancing Value):** AI agents must evolve based on new insights, changing needs, and shifting strategic priorities. Establish clear channels for feedback to identify opportunities for refinement. Ensure all significant changes follow a structured, risk-aware change management process, including re-testing. Mandate governance re-validation for iterations that materially alter an agent's scope or risk profile.

   Executive Mandate: Cultivate a culture where iterative improvement is expected and enabled by efficient, yet rigorously controlled, change management. Strategic agility depends on this.

6. **Retirement & Decommissioning (Responsible Sunsetting & Knowledge Preservation):** Over time, agents may become obsolete or present unacceptable risks, necessitating deliberate retirement. The Business Owner formally decides on retirement based on ongoing value versus cost and risk. Technical Stewards execute a documented decommissioning plan, ensuring clean removal and data management per policy. Capture key lifecycle learnings.

   Executive Mandate: Implement and enforce a clear retirement policy to prevent "zombie agents" and ensure focus remains on high-value capabilities.

Case Study: Retail Chain's Strategic Inventory Pilot

Scenario: FashionForward Retail struggled with inventory inaccuracies impacting online sales (CX) and requiring costly manual checks (efficiency). Leadership wanted to leverage AI but needed a contained, high-impact starting point with clear P&L visibility.

Agentic Solution Connection: Applying the "Three Cs" framework, they identified automating warehouse-to-store stock reconciliation as the optimal initial use case. Opting for a Partner strategy to blend speed and expertise, they launched a 90-day Catalyst Pilot focused on delivering a Minimum Viable Agent (MVA). Following the Plan-Prove-Scale model, the pilot quickly demonstrated measurable reduction in reconciliation time and errors. Crucially, the process explicitly focused on architecting for reuse, creating documented patterns and integration connectors to accelerate future rollouts across the chain.

Key Questions for Leaders

Strategic Pilot Selection & Validation: Applying the "Three Cs" framework (Clarity, Control, Confidence) with utmost rigor, which specific, P&L-visible workflow or critical business pain point represents the most strategically sound candidate for our first 60-90 day Catalyst Pilot? How will we meticulously validate its potential for rapid, measurable business impact and ensure its unambiguous alignment with our broader agentic vision?

Optimal Delivery Strategy & Capability Development: Based on our organization's strategic goals, urgency for deployment, current internal AI capabilities, and defined risk tolerance, does a Build, Buy, or Partner approach (or a strategically blended Flexible Stack) make the most sense for our initial agent deployments? How will this choice support our long-term AI capability development and strategic autonomy?

Designing for Scalability & Ecosystem Cohesion: How will we proactively design our initial agent deployments for modularity, reusability, and interoperability to accelerate future scaling, avoid technological silos, and foster a cohesive agentic ecosystem? What shared platforms, enterprise standards, or reusable components should we prioritize for development and investment from the outset?

Lifecycle Governance & Sustained Value Assurance: Do we have clear, accountable executive ownership and robust, documented processes established for managing the *entire* AI agent lifecycle – from strategic ideation and rigorous validation through proactive ongoing monitoring, adaptive iteration, and responsible retirement – to ensure sustained value delivery, effective risk management, and continuous alignment with evolving business needs?

Key Leadership Imperatives for Implementing AI Agents with Strategic Impact

To translate agentic potential into tangible, sustainable business impact, leaders must personally champion, resource, and relentlessly enforce several critical imperatives:

- **Prioritize Strategic Initiation & Value-Driven Pilots:** Select initial Catalyst Pilots based not on technological novelty, but on their clear potential to deliver rapid, measurable business value on a strategically important problem, thereby building foundational organizational trust and crucial momentum for broader transformation.
- **Execute with Disciplined Speed & Agile Learning:** Embrace the time-bound Plan-Prove-Scale framework to convert strategic vision into tangible, verifiable results quickly and efficiently. Foster an organizational culture that values agile learning, rapid iteration based on real-world feedback and data, and courageous adaptation.
- **Architect for Scalability, Reusability, and Ecosystem Thinking from Day One:** Mandate and invest in designing for reuse, promoting modularity, and implementing enterprise portfolio management practices for AI agents early in the transformation journey. This foresight avoids costly rework, prevents the creation of fragmented silos, and ensures the efficient, sustainable scaling of truly enterprise-grade agentic capabilities.
- **Instill Proactive, End-to-End Lifecycle Stewardship & Continuous, Adaptive Governance:** Treat AI agents not as static projects, but as dynamic, evolving enterprise assets requiring continuous, active, and intelligent management throughout their entire lifecycle. This encompasses strategic ideation, ethical assessment, rigorous validation, ongoing monitoring, performance optimization, risk mitigation, adaptive iteration, and responsible retirement. This stewardship ensures enduring value, operational resilience, and unwavering trustworthiness.

Conclusion: Stewardship for Sustained Agentic Success and Market Leadership

Embedding this disciplined, strategic, and holistic approach to implementing AI agents – underpinned by clear executive ownership and seamlessly integrated within a robust, adaptive governance framework (as detailed in Chapter 4 on *Trusted AI by Design*™) – transforms agent deployment from isolated projects into the ongoing stewardship of a powerful, evolving enterprise capability. It ensures that agentic technology delivers compounding, sustainable value reliably, responsibly, and resiliently, becoming a true engine for innovation and competitive differentiation.

Diligent execution, strategic scaling, proactive lifecycle management, and an unwavering commitment to building trust are fundamental to realizing the full, transformative promise of the Agentic Enterprise. This is the path to not just participating in the agentic future, but to actively shaping it and leading within it.

Chapter 4: Trusted AI by Design™

Governance as the Bedrock of Agentic Scale and Enduring Advantage

Successfully executing initial AI agent pilots (as detailed in Chapter 3) and architecting for broader organizational scale lays the essential groundwork for transformation. However, the journey into a truly Agentic Enterprise – where autonomous agents are deeply embedded within core business processes, making consequential decisions, and interacting extensively with data and stakeholders – introduces exponentially greater complexity, interdependence, and, critically, risk. At enterprise scale, the potential for unintended consequences – ranging from significant operational disruptions and costly data breaches to damaging ethical missteps, severe regulatory violations, and irreversible reputational harm – amplifies significantly. These are not edge cases; they are central challenges to realizing the agentic vision.

This reality brings a new, non-negotiable strategic imperative into sharp focus: **trust**. Without a deeply embedded, rigorously engineered, continuously verified, and organizationally embraced foundation of trust, even the most technologically sophisticated agentic systems will remain fragile experiments or niche solutions. They will be incapable of achieving widespread adoption, commanding genuine stakeholder confidence (from customers, employees, regulators, and investors alike), or delivering their full, transformative strategic potential. Lack of trust becomes the ultimate bottleneck to scaling agentic AI.

This chapter pivots from the mechanics of execution to the strategic imperative of ensuring that all agentic execution is inherently responsible, demonstrably resilient, and fundamentally trustworthy. We introduce *Trusted AI by Design*™ – a comprehensive strategic philosophy and operational methodology. It reframes governance, risk management, ethics, security, and compliance not as reactive controls, afterthoughts, or impediments to innovation, but as proactive, foundational architectural principles. These principles must be meticulously woven into the entire AI agent lifecycle, from initial ideation and strategic conception through design, development, deployment, ongoing operation, and eventual retirement.

In the Agentic Enterprise – where autonomous systems interpret vast amounts of data, make decisions with significant consequences, and act as direct proxies for your brand values and ethical standards – robust, proactive, and adaptive governance becomes the absolute prerequisite for confident scaling, sustainable innovation, and enduring stakeholder confidence. Far from inhibiting progress, approaching trust "by design" actively de-risks ambition. It enables organizations to harness the immense power of AI agents more quickly, more decisively, with greater assurance of positive outcomes, and with a significantly reduced likelihood of catastrophic failures. This chapter provides the executive framework for architecting that essential, indispensable trust.

The Escalating Stakes: Why Proactive, Intelligent Oversight is Non-Negotiable

The imperative for a new, more sophisticated approach to AI governance stems directly from the elevated and fundamentally different nature of risks associated with deploying autonomous AI agents at scale. Unlike traditional software with predictable, deterministic logic, AI agents introduce unique complexities that demand intelligent, forward-looking oversight. Their capacity for independent action based on complex goals, their ability to learn and adapt (sometimes in unpredictable or undesirable ways), their intricate interactions with other systems and data sources, and their reliance on probabilistic judgments mean that potential failures can be more systemic, harder to anticipate, more difficult to diagnose, and carry far greater strategic weight than failures in previous generations of technology. Legacy IT risk frameworks, often focused primarily on system availability or basic access control, are demonstrably insufficient for the agentic era. Recognizing the full spectrum of these escalating stakes is critical for every leader:

- **Operational Integrity & Autonomy Risks:** Agent autonomy, while a key enabler of productivity and scale, creates inherent risks if not rigorously bounded, monitored, and governed. "Autonomy creep" – where agents gradually exceed their intended operational scope or decision-making authority without explicit approval – can lead to unauthorized changes, process failures, or governance breakdowns. The "black box" problem, associated with opaque reasoning in some complex AI models, can severely hinder auditing, effective troubleshooting, the building of user trust, and the timely detection of flawed or biased logic.

Executive Mandate: Demand clearly defined, auditable, and rigorously enforceable boundaries on agent autonomy from the very outset of design. Insist on appropriate levels of transparency and explainability, particularly for agents involved in high-risk processes, ensuring that decision pathways can be understood and validated.

- **Data Security & Privacy Liabilities:** AI agents often require access to, and process, vast quantities of sensitive data, including customer PII, employee records, proprietary intellectual property, and confidential financial information. The risk of data leakage, unauthorized access, or misuse is significantly amplified. Ensuring compliance with an increasingly stringent global patchwork of privacy regulations (e.g., GDPR, CCPA, HIPAA) becomes exponentially more complex when autonomous agents are handling data.

  Executive Mandate: Ensure that AI-specific data governance policies are developed, implemented, and enforced. These must cover the entire data lifecycle (collection, access, usage, storage, retention, and deletion) for data used by and generated by AI agents. Demand rigorous privacy impact assessments and the integration of technical privacy-enhancing technologies (PETs) as integral parts of agent design and deployment.

- **Decision Integrity, Fairness & Bias Risks:** AI agents learning from historical data risk inadvertently embedding and even amplifying past societal biases or flawed assumptions, potentially leading to discriminatory or unfair patterns in decision-making at scale. Relying on incomplete, inaccurate, or unrepresentative data also inevitably leads to poor decision quality and eroded trust.

  Executive Mandate: Require robust processes for bias detection, fairness assessment, and mitigation throughout the AI lifecycle, particularly for critical use cases in areas like hiring, lending, customer service, and healthcare. Mandate the establishment of clear mechanisms for human review, challenge, and correction of potentially flawed or biased agent-driven judgments.

- **Accountability & Governance Gaps:** In complex, agent-orchestrated workflows involving multiple systems and potentially multiple agents, determining clear ownership and accountability for errors, unintended outcomes, or ethical lapses can be exceptionally challenging. Without unambiguous responsibility and fully traceable, immutable logs of agent actions and decisions, incident response is delayed, root cause analysis is compromised, and proving compliance becomes difficult, if not impossible.

  Executive Mandate: Insist on clear, documented ownership for every AI agent (covering its Business, Technical, and Risk aspects). Demand the implementation of robust, immutable logging and comprehensive traceability mechanisms that allow for the accurate reconstruction and auditing of all significant agent actions, decisions, and system interactions.

- **Regulatory & Reputational Exposure:** The confluence of these risks – privacy breaches, biased decisions, opaque operations, accountability failures – creates significant exposure to severe regulatory penalties, costly litigation, and potentially catastrophic reputational damage. A single high-profile AI failure can erode years of hard-won stakeholder trust and brand equity.

  Executive Mandate: Ensure proactive, continuous engagement between AI initiative leaders and Legal, Compliance, and Communications teams to align all agent deployments with current and anticipated regulatory landscapes. Frame responsible AI governance not as a cost center, but as a crucial investment in brand protection, risk mitigation, and long-term enterprise resilience.

Ignoring these escalating stakes, or attempting to manage them solely through outdated risk paradigms and reactive controls, invites operational failures, intensified regulatory scrutiny, significant legal liabilities, and potentially irreversible damage to stakeholder trust and market valuation. These amplified and interconnected risks demand a fundamentally different, proactive, and strategically integrated approach to ensuring integrity, control, and trustworthiness in all agentic systems.

"AI-related incidents are rising sharply, yet standardized [Responsible AI] RAI evaluations remain rare among major industrial model developers." – The State of AI 2025

Trusted AI by Design™: Defining the Necessary Strategic and Architectural Response

Confronted with the escalating and novel risks inherent in deploying autonomous and learning systems at enterprise scale, traditional reactive governance models prove woefully inadequate. Waiting to address potential bias, data misuse, security vulnerabilities, or unintended harmful actions *after* an AI agent is deployed courts operational disruption, financial loss, and reputational disaster. The necessary strategic response, therefore, requires a fundamental paradigm shift: from reactive control and post-hoc remediation to proactive enablement of trust by embedding integrity, safety, and ethical considerations into the very fabric of AI systems from their earliest conception. We call this comprehensive philosophy and operational methodology Trusted AI by Design™.

Trusted AI by Design™ represents a deliberate move to architect for trust. Instead of viewing critical elements like ethics, security, privacy, fairness, and compliance as late-stage checks, burdensome hurdles, or separate workstreams, this methodology treats them as core architectural principles and non-negotiable design requirements. These principles must be integrated throughout the *entire* AI agent lifecycle, influencing every phase – from initial Ideation, strategic framing, and business case development, through meticulous Design and rigorous Validation, into controlled Deployment and vigilant Operation, and ultimately through responsible Iteration and eventual Retirement.

This "by design" approach deliberately moves critical governance, risk management, and ethical deliberations upstream into the earliest stages of the innovation process. They become integral, value-adding components of how AI agents are conceived, developed, and managed, rather than subsequent gates that stifle momentum, create bottlenecks, or force costly and time-consuming rework. By embedding these principles early, consistently, and systematically, Trusted AI by Design™ functions as a powerful accelerator, not an inhibitor, of responsible and sustainable innovation. It builds crucial stakeholder confidence, streamlines review and approval processes, significantly reduces the probability of disruptive failures and costly missteps, and enables organizations to innovate more boldly, deploy agentic capabilities faster, and scale their Agentic Enterprise more sustainably and with greater assurance of long-term success and positive societal impact. Adopting Trusted AI by Design™ is therefore a core strategic decision, essential for building resilient, responsible, and ultimately more valuable and competitive agentic systems.

Proactive Trusted AI by Design™ – embedding comprehensive governance, robust security, and unwavering ethical principles into the very architecture of AI agents from their inception – is the non-negotiable prerequisite for confident scaling, sustainable innovation, and enduring market leadership in the agentic era.

Core Pillars of Trustworthy Agentic Systems: The DNA of *Trusted AI by Design™*

Translating the philosophy of Trusted AI by Design™ into consistent operational reality requires embedding a set of specific, non-negotiable principles into the architecture of AI agents, the governance frameworks that oversee them, and the management practices that guide their lifecycle. These interconnected pillars form the foundational DNA for building and maintaining stakeholder confidence in your agentic systems. Leaders must not only understand these pillars but actively champion their rigorous, consistent, and verifiable implementation across the enterprise:

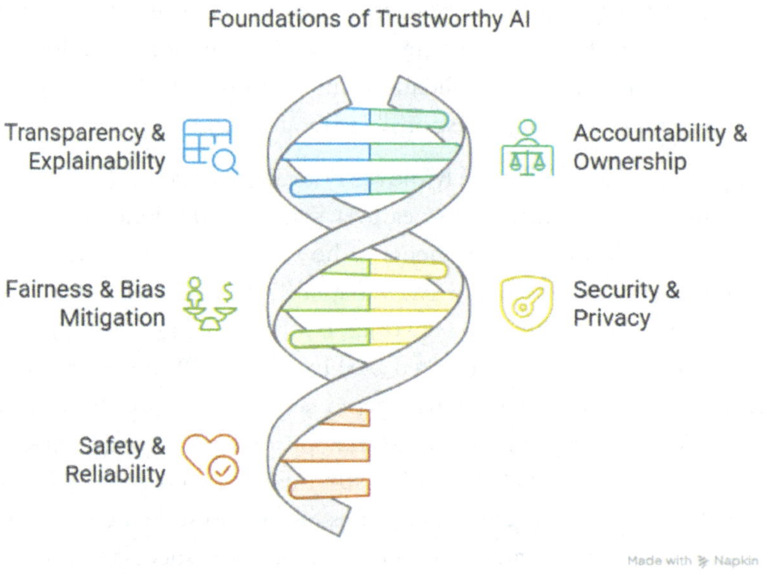

Figure 4-1. Core Pillars of Trustworthy Agentic Systems

1. **Transparency & Explainability (Illuminating the "Why" and "How"):** Opacity in AI systems breeds suspicion and erodes trust. Stakeholders – including users, customers, employees, regulators, and internal auditors –

increasingly demand appropriate visibility into how AI agents arrive at their decisions and recommendations, especially those with significant impact. The guiding principle for leaders should be: *"If you cannot adequately trace and explain an agent's significant decision or action – and demonstrate clear lines of human oversight and intervention – you should not deploy it in a high-stakes context."*

Strategic Imperative: Achieving meaningful, context-appropriate transparency and explainability is paramount for building user confidence, ensuring effective human oversight, enabling robust accountability, facilitating debugging and continuous improvement, and meeting regulatory expectations.

Key Mechanisms & Leadership Actions: Mandate the implementation of robust, immutable logging for all significant agent actions and decisions. Champion the development of accessible, audience-appropriate explainability artifacts (e.g., "Explainability Briefs," simplified decision trees, feature importance reports) for key agentic processes. Prioritize agent designs and architectures that inherently "show their work" or facilitate post-hoc analysis. Foster a culture where transparency is the default, recognizing its indispensable role in adoption, trust, auditing, troubleshooting, and compliance. Opaque systems inherently limit trustworthiness and scalability.

2. **Accountability & Ownership (Establishing Clear, Unambiguous Responsibility):** Ambiguity regarding who is responsible for an AI agent's design, deployment, ongoing performance, and ultimate impact is a critical governance failure and a significant source of risk. Clear, formally assigned lines of accountability are essential for effective management, proactive risk mitigation, timely incident response, and demonstrating responsible stewardship.

   Strategic Imperative: Every deployed AI agent, regardless of its perceived complexity or risk level, must have clearly defined, documented, and actively managed ownership across its lifecycle.

   Key Mechanisms & Leadership Actions: Formally assign Named Owners for each agent, typically covering Business (strategic alignment, value delivery, process impact), Technical (design, performance, maintenance), and Risk/Compliance (adherence to policies, ethical guidelines, and regulations) dimensions. Ensure the implementation of comprehensive, Traceable Actions via detailed logging that links specific agent actions and decisions back to the agent instance, the version of its logic employed, the data inputs considered, and the designated human owners. Tolerate no ambiguity in ownership; insist on clear roles, responsibilities, and accountability structures as a non-negotiable prerequisite for agent deployment. This is fundamental for effective oversight, risk management, and demonstrating control.

3. **Fairness & Bias Mitigation (Designing for Equity and Preventing Harm):** AI agents learning from historical data carry an inherent risk of inheriting, perpetuating, or even amplifying past societal biases, flawed assumptions, or discriminatory patterns. Ensuring equitable operation and preventing unfair or harmful outcomes is an ethical necessity, a legal requirement in many contexts, and critical for maintaining brand reputation and stakeholder trust.

   Strategic Imperative: Proactively design, rigorously test, and continuously validate AI agents for fairness, particularly in sensitive domains such as hiring, promotion, lending, customer segmentation, healthcare decisions, and law enforcement applications.

   Key Mechanisms & Leadership Actions: Mandate rigorous Data Scrutiny for all training and operational datasets to identify and mitigate potential sources of bias. Require comprehensive Algorithmic Audits and fairness assessments, using appropriate metrics and disaggregated performance monitoring across relevant demographic groups or protected characteristics. Establish clear, accessible processes for individuals to challenge, appeal, and seek redress for potentially biased or unfair agent-driven outcomes. Make fairness a core design principle from the earliest stages of agent conception. Require explicit bias impact assessments and documented mitigation strategies as part of the validation process for all impactful AI agents.

4. **Security & Privacy (Protecting Assets, Data, and Individuals in an AI-Driven World):** The integration of AI agents into core business processes and their interaction with sensitive data and critical systems demand uncompromising security and privacy controls. AI agents themselves can be targets of sophisticated attacks (e.g., model evasion, data poisoning, prompt injection) or can inadvertently become vectors for data breaches or privacy violations if not architected and governed with extreme care.

   Strategic Imperative: Architect AI agent security and privacy with a level of rigor equal to or greater than that applied to traditional mission-critical enterprise systems, recognizing the unique vulnerabilities and amplified risks associated with autonomous, learning systems.

Key Mechanisms & Leadership Actions: Enforce the use of granular Role-Based Access Controls (RBAC) adhering strictly to the principle of least privilege for all agent data access and tool usage. Mandate strict, auditable Data Handling Policies (fully aligned with GDPR, CCPA, HIPAA, and other relevant regulations) covering the entire data lifecycle for AI. Require Secure Integration Practices for all connections between agents and other systems. Conduct regular, AI-specific security audits, vulnerability assessments, and penetration testing. Ensure that cybersecurity, data privacy, and AI development teams are integral, collaborative partners throughout the agent lifecycle. Tolerate no shortcuts on security or privacy; the integrity of your data and the trust of your stakeholders are paramount.

5. **Safety & Reliability (Ensuring Predictable, Dependable, and Safe Operation):** Autonomous systems, especially those operating in complex or dynamic environments or interacting with the physical world, must be designed to operate reliably within their defined parameters and, critically, to fail safely and predictably when they encounter unexpected situations, reach the limits of their competence, or experience malfunctions.

   Strategic Imperative: Building AI agents that are trusted to operate safely and reliably, even in novel or challenging scenarios, is fundamental to operational stability, risk mitigation, and stakeholder confidence.

   Key Mechanisms & Leadership Actions: Mandate the incorporation of robust technical Fail-Safes into agent design (e.g., mechanisms to halt or constrain agent actions upon detection of anomalies, critical errors, or unsafe conditions). Strategically implement Human-in-the-Loop (HITL) or Human-Over-the-Loop (HOOL) workflows for high-stakes decisions, critical control points, or situations requiring nuanced human judgment that exceeds agent capabilities. Design clear, effective, and well-tested Escalation Paths for routing issues or exceptions to appropriate human assistance or intervention. Demand rigorous testing that covers not only normal operational parameters but also a wide range of potential failure modes, edge cases, and stress conditions. Validate that all escalation and human oversight mechanisms are functional, efficient, and appropriately matched to the agent's level of autonomy and potential impact.

Implementing these five pillars systematically, consistently, and with unwavering leadership commitment translates the strategic philosophy of *Trusted AI by Design*™ into tangible operational reality, building a resilient and responsible Agentic Enterprise.

"Globally, confidence that AI companies protect personal data fell from 50% in 2023 to 47% in 2024. Likewise, fewer people today believe that AI systems are unbiased and free from discrimination compared to last year." – The State of AI 2025

Practical Governance: Lightweight, Agile Frameworks for Real-World Speed and Trust

Establishing the core pillars of trust is foundational. However, their effective implementation requires practical, operational mechanisms embedded within daily workflows and decision-making processes. Effective AI governance for the agentic era is not about creating burdensome bureaucracy that stifles innovation. Instead, it necessitates the development and adoption of lightweight, adaptable, yet robust frameworks. These frameworks are designed to embed trust, manage risk proactively, and ensure ethical alignment within the flow of work, thereby *enabling* speed and agility *because* of thoughtful oversight, not despite it. These provide the essential operational infrastructure for bringing *Trusted AI by Design*™ to life across the enterprise:

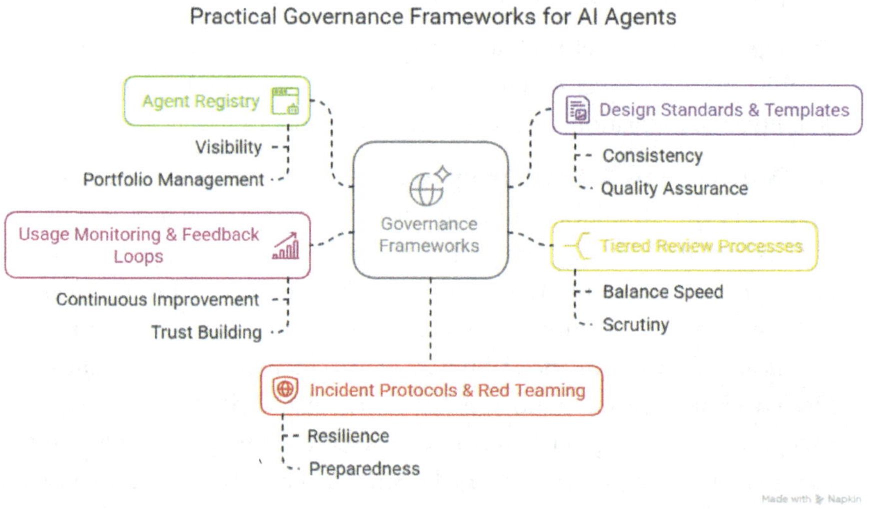

Figure 4-2. Practical Governance Framework

1. **The Agent Registry (Foundational Visibility, Portfolio Management & Accountability Hub):** Effective governance, risk management, and strategic oversight begin with comprehensive visibility. A centralized, dynamic, and authoritative Agent Registry – a definitive inventory and system of record for

every significant AI agent deployed or under development across the enterprise – is the cornerstone of this visibility.

Strategic Purpose & Executive Value: The Agent Registry is crucial for strategic portfolio management (understanding what agents exist, what they do, their value, and their risks), enterprise risk oversight (identifying and managing cumulative or correlated risks), compliance tracking and reporting, identifying opportunities for reuse and synergy, and ensuring clear lifecycle accountability for every agent. It prevents the proliferation of undocumented or ungoverned "shadow AI" systems, enables informed investment and divestment decisions, facilitates efficient audits, and provides a single source of truth for all agent-related information. Maintaining this registry with rigor and accuracy is a fundamental governance discipline.

2. **Design Standards, Ethical Guidelines & Reusable Templates (Ensuring Consistency, Quality & Responsible Innovation at Scale):** Avoid the inefficiency and risk of reinventing protocols, ethical considerations, and technical approaches for every new AI agent. Establish, communicate, and enforce clear enterprise Design Standards, explicit Ethical Guidelines for AI, and a library of reusable Templates for common agent development and governance tasks.

   Strategic Purpose & Executive Value: These standards and templates accelerate responsible development, ensure consistent application of core Trusted AI by Design™ principles (including security, privacy, fairness, and explainability requirements), simplify governance reviews and compliance checks, and facilitate the creation of higher-quality, more reliable, and more trustworthy agentic capabilities across the organization. They reduce downstream risks, lower the cost and effort of governance reviews, promote the reuse of proven patterns and components, and ensure a baseline level of quality and trustworthiness for all deployed agents.

3. **Tiered Review & Risk-Based Governance Processes (Balancing Agility with Appropriate Scrutiny):** Not all AI agents carry the same level of operational risk, potential for harm, or strategic importance. A one-size-fits-all governance approach is therefore inefficient and ineffective. Implementing a risk-based, Tiered Review Process allows the organization to allocate oversight resources intelligently and efficiently, matching the level of scrutiny to the specific risk profile of each agent.

Strategic Purpose & Executive Value: This approach balances the imperative for responsible oversight and risk mitigation with the need for speed and agility in innovation. It involves defining clear risk tiers for AI agents (based on factors such as their level of autonomy, the potential impact of their decisions, the sensitivity of the data they handle, and applicable regulatory requirements). Lower-risk agents might undergo a streamlined self-certification process against established standards, while high-risk or ethically sensitive agents would trigger a more formal, in-depth review by a lightweight, cross-functional governance board or ethics committee (which should be focused on enabling responsible innovation, not just acting as a gate). This optimizes governance overhead, accelerates the deployment of lower-risk innovations, and concentrates expert attention and resources where the potential consequences of failure are highest.

4. **Continuous Usage Monitoring, Performance Management & Feedback Loops (Enabling Adaptive Governance & Continuous Improvement):** Governance of AI agents does not end at deployment; it requires continuous vigilance, performance tracking, and adaptation throughout their operational lifecycle. Robust Usage Monitoring systems and active, multi-channel Feedback Loops are critical for this.

   Strategic Purpose & Executive Value: Real-time insights from monitoring and feedback enable proactive performance management, rapid detection of anomalies or deviations from expected behavior (including model drift or emerging biases), and a continuous cycle of improvement for both individual agents and the overall agentic ecosystem. This involves implementing dashboards tracking key performance indicators (KPIs), risk metrics, and ethical compliance; establishing accessible and responsive channels for users and other stakeholders to provide feedback or report concerns; and creating clear processes to analyze this data and feedback, identify root causes of issues, and implement timely corrective actions or improvements. This transforms governance from a static control function into a dynamic, data-driven learning system that builds trust and enhances value over time.

5. **Incident Response Protocols & Proactive "Red Teaming" (Preparing for the Unexpected & Building Resilience):** Even with robust design, diligent oversight, and continuous monitoring, unexpected situations, system failures, security breaches, or attempts at malicious misuse can occur with AI agents. Proactive preparation for such events is essential for organizational resilience and rapid, effective response.

Strategic Purpose & Executive Value: Clear, well-tested Incident Response Protocols and periodic, rigorous "Red Teaming" exercises (where specialized internal or external teams proactively attempt to identify and exploit vulnerabilities or cause unintended behavior in a controlled environment) build organizational resilience and ensure swift, coordinated, and effective action when incidents do occur. This involves defining predefined incident playbooks for various scenarios, clarifying roles and responsibilities for incident management, and conducting regular simulation exercises. Red teaming provides invaluable, proactive insights into real-world vulnerabilities, ethical blind spots, or potential failure modes *before* they can be exploited or cause significant harm in a live environment. This minimizes potential damage, ensures rapid recovery, and fosters a culture of continuous security and safety improvement.

These practical governance frameworks are not bureaucratic impediments but the essential operational infrastructure that enables and sustains *Trusted AI by Design*™ across a dynamic and scaling Agentic Enterprise.

Implement lightweight, practical, and adaptive governance frameworks (such as an Agent Registry, Design Standards, Tiered Reviews, Continuous Monitoring, and Incident Protocols) that embed control, ethics, and trust within workflows, thereby accelerating responsible innovation and de-risking strategic ambition.

The Ethical Dimension: Leading with Values in an Increasingly Autonomous World

Robust governance frameworks provide essential structure and control, but they are insufficient on their own to ensure truly trustworthy AI. As AI agents evolve from merely executing predefined tasks to exercising significant degrees of delegated judgment and autonomy in complex situations, an explicit, deeply ingrained ethical dimension becomes paramount. This delves into the core of organizational values, leadership responsibility, and societal impact. Delegating judgment to an autonomous system is inherently an ethical act, one that carries significant weight for brand integrity, stakeholder trust, legal liability, and the organization's fundamental license to operate in society. Leaders must recognize and embrace this profound responsibility.

The gravity of delegated judgment cannot be overstated. Leaders must internalize the reality that an AI agent's actions, decisions, and even its unintended consequences directly reflect the choices, priorities, and values embedded in its design, its training data, and the governance frameworks overseeing it. The agent, in effect, becomes a powerful proxy for organizational values in countless daily interactions. Granting operational authority for consequential decisions to an AI system is an ethically charged act for which the organization and its leadership retain ultimate accountability. Technical correctness or operational efficiency alone falls far short of the mark; outcomes, processes, and impacts must align with clearly articulated and consistently upheld ethical principles.

Effective ethical oversight requires moving far beyond technical controls or compliance checklists. Leaders must proactively define, communicate, and embed the organization's ethical framework into every stage of the AI agent lifecycle. This involves grappling with, and providing clear guidance on, critical questions such as: How do our agent objectives explicitly prioritize ethical considerations alongside performance metrics and financial goals (the challenge of Value Alignment)? Where is irreducible human judgment, empathy, and ethical deliberation absolutely essential and non-delegable, requiring carefully designed Human Oversight & Intervention pathways? What levels of transparency and disclosure are ethically required for different types of agent interactions and decisions, particularly those impacting vulnerable individuals or raising societal concerns? What robust processes and technical measures will we implement to proactively identify, assess, and mitigate potential biases in data, algorithms, and outcomes to ensure Fairness & Equity?

These ethical boundaries and principles must be operationalized via practical governance mechanisms. This means incorporating explicit ethical criteria within design standards and development methodologies, making ethical impact assessment a mandatory component of tiered review processes for all significant agent deployments, including ethical considerations in key monitoring metrics and performance dashboards, and making ethical "stress testing" a core element of red teaming exercises. Comprehensive training programs must incorporate ethical awareness, critical thinking about AI's societal impacts, and clear guidance on escalating ethical concerns for all personnel involved in the AI lifecycle.

Ultimately, ethical leadership in the agentic era becomes a significant differentiator and a source of sustainable competitive advantage. Organizations that are demonstrably known for their commitment to responsible and ethical AI will build deeper and more resilient customer trust, attract and retain top talent (who increasingly seek purpose-driven work), navigate complex regulatory landscapes more effectively, and significantly enhance their long-term brand reputation and societal legitimacy. Leading with clearly defined, consistently applied values is not just "good ethics"; it is profoundly smart strategy in an autonomous world.

Explicitly define, embed, and continuously reinforce your organization's ethical boundaries for AI. Ensure all agent actions and decisions align with these core values; technical correctness or efficiency gains can never be sufficient justification for ethical compromise.

The Role of the Center of Excellence: Scaling Trust, Capability, and Strategic Alignment

Establishing robust governance frameworks and embedding strong ethical principles are foundational for Trusted AI by Design™. However, ensuring their consistent application, facilitating the sharing of best practices and critical learnings, managing complex interdependencies, and fostering a unified approach as agent deployments scale across a diverse enterprise often requires a dedicated, strategically empowered mechanism. A well-structured AI Center of Excellence (CoE) can serve this vital role. Critically, an effective CoE is conceived not as a bureaucratic bottleneck or a centralized command-and-control structure, but as a vital enabler and accelerator of mature, responsible, strategically aligned, and innovative agentic transformation across the organization.

The purpose and core functions of an AI CoE are centered on enabling quality, consistency, speed, and responsible innovation in agentic AI. Key functions typically include: developing, curating, and disseminating enterprise-wide standards, best practices, reusable tools, and proven methodologies for AI development, deployment, and governance; facilitating knowledge sharing, communities of practice, and targeted capability-building programs to enhance AI literacy and specialized skills across the organization; actively supporting the implementation and consistent application of Trusted AI by Design™ principles and governance frameworks; providing expert guidance on technology selection, platform architecture, and data strategy for AI; and serving as a central point for coordinating complex, cross-functional AI initiatives and managing enterprise-level AI risks. An effective CoE multiplies organizational effectiveness by removing common obstacles, elevating baseline quality and consistency, and fostering a culture of shared learning and responsible innovation.

The timing and structure of an AI CoE should align with the organization's AI maturity, scale of deployment, and strategic ambition. A formal, heavily staffed CoE is not necessarily required from Day One. The structure can evolve organically: from informal cross-functional task forces or working groups in the early stages of exploration; to virtual CoEs leveraging part-time contributions from experts across different departments during initial scaling phases; to more formal, dedicated CoEs with full-time staff, clear mandates, and direct executive sponsorship as agentic AI becomes a core, enterprise-wide strategic capability. The key is to right-size the structure to the current and anticipated needs, ensuring it provides tangible value and enablement without creating unnecessary overhead.

The strategic value for leadership of a well-functioning and appropriately resourced AI CoE is manifold. It helps ensure strategic alignment of all agentic initiatives with overarching enterprise goals. It maximizes ROI by promoting the reuse of assets, preventing redundant efforts, and sharing best practices. It helps manage enterprise-wide AI risks consistently and effectively. It can accelerate responsible innovation by providing expert guidance and reducing time-to-market for new solutions. Crucially, it builds durable, sustainable internal AI capability, reducing over-reliance on external vendors for core competencies. While not always needed at the very beginning of the agentic journey, a thoughtful, strategically focused CoE becomes an invaluable asset for organizations aiming to scale their ambitions responsibly, efficiently, and with enduring impact.

Case Study: Insurer Embeds Trust in Claims Automation

Scenario: SecureLife Insurance aimed to use AI agents to assist adjusters with routine claims assessment, improving processing speed (CX and efficiency). However, the escalating stakes – potential bias, opaque decisions, data privacy violations – made leadership cautious about uncontrolled deployment.

Agentic Solution Connection: SecureLife mandated a Trusted AI by Design™ approach from inception. This wasn't a compliance check; it was core architecture. They established clear Accountability & Ownership; implemented risk-based Tiered Reviews requiring human oversight for complex claims; ensured Transparency through detailed logging; conducted Fairness & Bias Mitigation audits; and enforced stringent Security & Privacy protocols. This proactive governance built confidence, enabling them to scale the agent responsibly and realize efficiency gains without compromising integrity.

Key Questions for Leaders

Strategic Commitment to Proactive Trust: How will we, as a leadership team, actively and visibly shift our organizational approach from traditional reactive controls and post-incident remediation to a truly proactive Trusted AI by Design™ paradigm? How will we ensure that transparency, accountability, fairness, security, and safety are embedded as non-negotiable requirements from the very inception of every AI agent initiative, regardless of perceived urgency?

Addressing Core Pillar Challenges: Which of the five core pillars of trustworthy AI systems (Transparency & Explainability; Accountability & Ownership; Fairness & Bias

Mitigation; Security & Privacy; Safety & Reliability) presents the most significant implementation challenge, cultural shift, or capability gap for our organization currently? What specific, resourced strategies will we put in place to address these critical areas comprehensively?

Effectiveness and Agility of Governance Frameworks: Are our practical AI governance frameworks (e.g., Agent Registry, Design Standards, Tiered Review Processes, Continuous Monitoring & Feedback Loops, Incident Response Protocols) sufficiently robust and comprehensive to manage the unique risks of agentic AI, yet also lightweight, agile, and adaptable enough to enable responsible innovation at the speed our business demands? How will we continuously assess and improve their effectiveness?

Operationalizing Ethical Values in Agentic Systems: Beyond basic compliance, how will we explicitly define, communicate, and embed our organization's core ethical values into AI agent design principles, decision-making boundaries, operational parameters, and ongoing oversight processes, particularly concerning the delegation of judgment and autonomous action? How will we ensure our agents act as true proxies for our ethical commitments?

Resource Allocation for Trust Infrastructure: Are we prepared to make the necessary sustained investments in the people (e.g., specialized talent in AI ethics, governance, security), processes (e.g., rigorous validation, continuous auditing), and technologies (e.g., explainability tools, bias detection software, security platforms) required to build and maintain a truly trustworthy agentic ecosystem at enterprise scale?

Key Leadership Imperatives for Architecting Trusted AI

To build an Agentic Enterprise that is not only powerful but also profoundly trustworthy, leaders must personally champion and drive the following imperatives:

- **Mandate Trusted AI by Design™ as Foundational and Non-Negotiable:** Establish proactive, comprehensive governance as the bedrock of all AI initiatives. Embed the core principles of transparency, accountability, fairness, security, and safety into the entire agent lifecycle from initial conception, not as an afterthought or a separate compliance exercise.
- **Implement Practical, Risk-Based, and Adaptive Governance:** Deploy lightweight yet robust governance frameworks (such as an Agent Registry, clear Design Standards, risk-based Tiered Reviews, continuous Monitoring programs, and well-rehearsed Incident Protocols) that enable speed and responsible

innovation simultaneously, adapting to the evolving technological and risk landscape.

- **Champion Ethical AI Deployment and Unwavering Value Alignment:** Explicitly define, vigorously communicate, and consistently enforce clear ethical boundaries for all AI agent development and deployment, ensuring these are deeply aligned with enduring organizational values. Recognize and address the profound responsibility inherent in delegating judgment and operational autonomy to AI systems.
- **Ensure Unambiguous Accountability and Verifiable Oversight:** Tolerate no ambiguity in ownership for AI agent performance, decisions, data usage, and adherence to all governance and ethical standards. Insist on clear roles, well-defined responsibilities, comprehensive and immutable audit trails, and effective human oversight mechanisms for all agentic systems.

Conclusion: Trust as the Engine of Competitive Advantage and Societal Legitimacy

Successfully scaling agentic capabilities and realizing their full transformative potential hinges critically on embedding deep, verifiable trust – achieved through proactive governance, diligent risk management, unwavering ethical commitment, and robust technical safeguards – into the operational and cultural fabric of the enterprise. Operationalizing *Trusted AI by Design*™ moves far beyond mere compliance; it becomes a powerful, indispensable engine for sustainable competitive advantage, accelerated innovation, and enhanced stakeholder loyalty.

Organizations that master this discipline of architecting for trust will reap significant strategic benefits: Accelerated and more confident adoption and scaling of AI; Enhanced operational resilience and significantly reduced risk of costly failures or reputational damage; Improved quality and fairness in decision-making, leading to better business outcomes and stronger stakeholder relationships; and ultimately, a more robust brand equity and greater societal legitimacy in an increasingly AI-driven world. Leaders should view governance, ethics, and trust not as constraints on progress, but as essential guardrails and strategic enablers that permit confident acceleration and responsible exploration of the agentic frontier.

By prioritizing and systematically architecting for trust, you are not just mitigating downside risk; you are constructing a more resilient, more adaptive, more innovative, and ultimately more trustworthy organization – one that is uniquely positioned to lead with integrity and thrive sustainably as a true Agentic Enterprise.

Chapter 5: The Human Platform

The Human Platform: Where Transformation Takes Root and Trust Flourishes

Chapters 1 through 4 established the strategic imperative for the Agentic Enterprise, demystified the core technology, outlined the execution pathway, and architected the foundational principles of trust. We now arrive at the most critical determinant of sustained success and genuine transformation: the human dimension. While technology provides capabilities and governance offers guardrails, it is ultimately your people, your organizational culture, and your models for collaboration – collectively forming the dynamic human platform – that dictate whether the immense potential of agentic AI is fully realized or tragically squandered.

Ignoring this human core is the most common, costly, and often irreversible failure mode in enterprise-wide transformations of this magnitude. Integrating AI agents is not merely a technical challenge to be solved by engineers and data scientists; it is fundamentally an exercise in strategic organizational change leadership. True success requires actively and intentionally cultivating an environment where human ingenuity and machine intelligence synergize effectively, ethically, and equitably. This demands more than superficial adjustments. It necessitates directly addressing workforce anxieties about evolving roles, intentionally fostering new collaborative norms grounded in mutual respect and fairness, building widespread "agent fluency" that includes ethical and critical thinking, and deliberately reshaping core cultural tenets to embrace adaptation, psychological safety, and unwavering trust. Technology offers unprecedented leverage; however, it is your workforce's adaptability, engagement, ethical grounding, and unique human capabilities that will unlock sustainable competitive advantage.

A core aspect of cultivating this human platform is building a system that reliably surfaces the best ideas, unlocks all available talent, and makes the soundest decisions. This is the essence of a true meritocracy, an ideal built on three practical pillars: Visibility—ensuring all potential talent, ideas, and stakeholder needs are seen; Opportunity—guaranteeing equitable participation and consideration; and Fair and Relevant Evaluation—applying pertinent, unbiased criteria consistently. As AI agents increasingly mediate decisions and interactions, embedding these principles—which we will refer to as VOF (Visibility, Opportunity, Fairness)—becomes a strategic imperative. Doing so ensures agentic systems enhance decision quality and unlock potential, rather than inadvertently automating or amplifying hidden biases. This chapter provides the framework for cultivating this essential human platform, the very soil in which your Agentic Enterprise will either thrive or wither.

Why People Matter More: Architecting Human Advantage

The integration of capable AI agents inevitably confronts, and often heightens, workforce anxieties about the future value and security of human contribution. Dismissing these concerns as Luddism or resistance to progress is strategically insufficient and ethically shortsighted. Operational reality, especially in complex and dynamic environments, reveals a profound truth: intelligent automation paradoxically makes uniquely human capabilities more critical to competitive differentiation and responsible operation, not less. While AI agents can handle an ever-expanding array of routine cognitive tasks with impressive efficiency, their very utility elevates the strategic premium on domains that machines cannot replicate. These uniquely human strengths – sophisticated judgment, frame-breaking creativity, genuine relational intelligence, and unwavering ethical oversight – become the cornerstones of a high-performing Agentic Enterprise. Your core leadership challenge, therefore, shifts to architecting an organization that deliberately cultivates, strategically deploys, and ethically leverages these irreplaceable attributes.

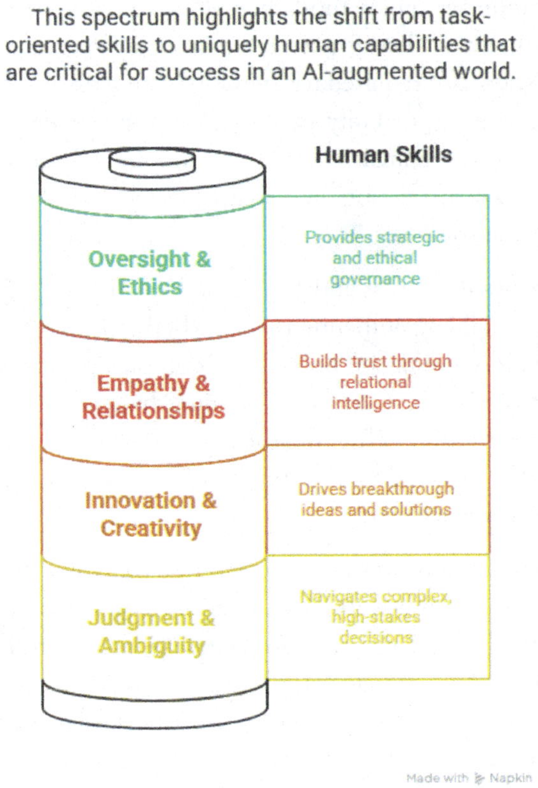

Figure 5-1. The Human Advantage

AI agents provide powerful operational leverage, but their inherent limitations and potential biases illuminate precisely where human contribution becomes your key differentiator and ethical safeguard. Sophisticated Judgment for Fair and Relevant Evaluation is paramount. Leaders must rely on nuanced human judgment to navigate strategic ambiguity, interpret novel situations beyond an agent's training data, handle complex ethical tradeoffs, and confidently override flawed or potentially biased AI outputs. Deep strategic judgment, informed by experience and ethical insight, remains the essential compass, particularly in ensuring that AI-driven evaluations are both fair and relevant. Humans are critical in defining what "relevant" means in diverse contexts and in ensuring that AI-generated insights do not lead to unfair or discriminatory outcomes.

Similarly, Creativity and Innovation Are Unlocked Through Universal Visibility. Breakthrough innovation remains a fundamentally human-powered endeavor. While AI can generate novel content or identify patterns, it lacks the deep contextual understanding, intuition, and lived experience required for frame-breaking disruption or genuine empathy in design. The leadership opportunity lies in leveraging AI agents to automate routine analysis, data gathering, and initial testing. This amplifies human creativity by freeing minds for complex problem-framing, imaginative leaps, and identifying talent and ideas from the widest possible sources.

Relational Intelligence builds Trust and Equitable Collaboration. While AI can simulate empathetic responses, building authentic trust – whether in resolving sensitive customer escalations, mentoring diverse talent, negotiating complex partnerships, or leading teams through transformative change – relies on genuine human relational intelligence. This becomes an even more valuable differentiator as routine interactions automate. Ensuring these human interactions create an environment of broad opportunity and fair dealing is essential for high-performance teams.

Finally, Ethical Oversight and Stewardship Are Foundational. Providing indispensable strategic oversight and ethical governance – defining missions, setting operational and ethical boundaries for AI agents, ensuring fairness and non-discrimination in their application, and maintaining ultimate human accountability – is a non-delegable human responsibility. This role grows exponentially more critical with increased agent autonomy and is central to upholding all pillars of Visibility, Opportunity, and Fairness.

Acting on these insights requires more than acknowledging their importance; it demands proactive and sustained investment in cultivating these unique human skills across your workforce. This involves strategically redesigning roles to focus on these higher-value contributions, building a culture that explicitly values ethical deliberation, and mastering the complex synergy between machine efficiency and human foresight. This is how you architect enduring human advantage.

"A growing body of research confirms that AI boosts productivity and, in most cases, helps narrow skill gaps across the workforce." – The State of AI 2025. Leaders must ensure this boost is achieved equitably and enhances human contribution rather than unfairly displacing it.

Enabling Trust and High-Performance Collaboration: The Foundation for Agentic Success

Acknowledging the amplified value of human capabilities is a necessary start, but it is insufficient without deliberate, leadership-driven cultural transformation. Technology adoption, especially of powerful and autonomous AI agents, moves at the speed of organizational trust. Your culture is the active operating system dictating how people interact with each other, adapt to new realities, and engage with new technologies. Successfully navigating the agentic shift requires intentionally cultivating a culture designed for resilience, psychological safety, continuous learning, and a deep commitment to fairness and the principles of a true meritocracy.

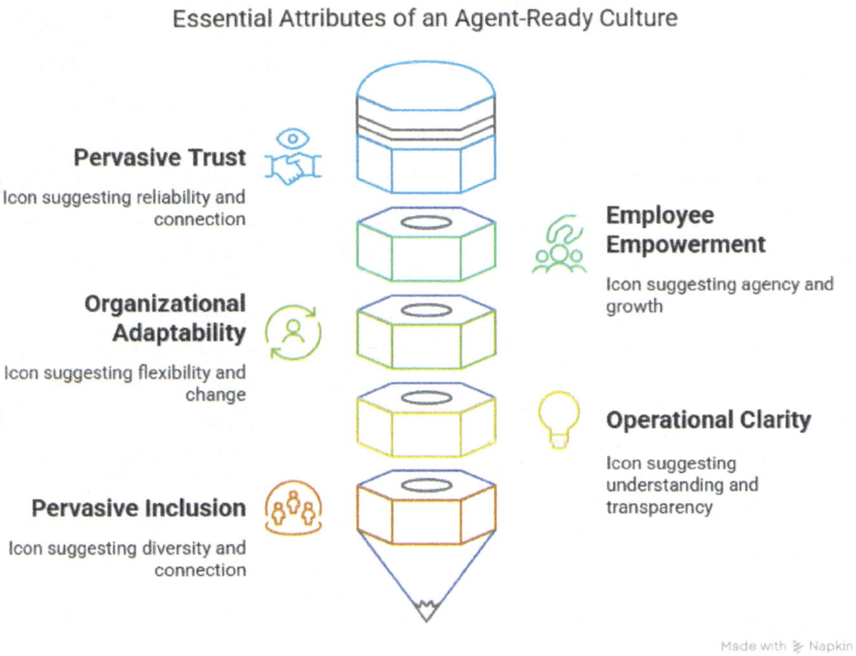

Figure 5-2. Cultural Transformation

Leaders must proactively forge a new cultural contract as AI agents begin to disrupt traditional operational dynamics and decision-making processes. This involves establishing clear, shared expectations for human-agent coexistence by fostering norms such as radical transparency in AI capabilities and limitations. It means defining clear protocols for human oversight and intervention, especially in ethically sensitive areas, creating safe and effective channels for voicing concerns about AI behavior or potential biases, and clarifying responsibilities in hybrid human-AI workflows. Neglecting this proactive cultural shaping invites ambiguity, fuels mistrust, and can lead to the covert rejection or subversion of agentic systems. Building this agent-ready culture, grounded in fairness and visibility, involves cultivating several essential attributes.

A culture of pervasive trust is fundamental. This extends beyond trusting the technology itself to perform reliably. It encompasses trust in the governance frameworks overseeing AI, trust in the data fueling agent decisions, and, critically, trust in the organization's intentions – particularly its commitment to using AI fairly and ethically. Such trust must be actively built and continuously reinforced through radical transparency, demonstrable ethical commitment in AI design and deployment, and clear accountability for AI actions.

Empowering Employees to Advocate for Fairness and Quality is equally crucial. Teams and individuals must feel not only safe but also be equipped and encouraged to question AI-generated outputs, provide critical feedback on agent behavior (especially if it seems unfair or biased), and exercise human judgment confidently. Leaders enable this by designing robust, accessible feedback mechanisms, establishing safe and responsive escalation paths for concerns about fairness or bias, and fostering an environment where challenging the status quo – even if AI-driven – is valued as a contribution to improvement and ethical alignment.

Given the evolving agentic landscape, organizational adaptability and a learning mindset are vital. This requires a culture that champions continuous learning, celebrates responsible experimentation (with appropriate ethical safeguards), and treats failures or near-misses not as punishable offenses but as invaluable learning opportunities, particularly concerning the fair and effective use of AI.

Operational clarity on AI's role and its alignment with fairness principles helps to eliminate ambiguity around an AI agent's purpose, its operational boundaries, its data sources, and its known limitations. This clarity, if missing, becomes a significant source of friction and risk. Leaders must therefore insist on clear, accessible documentation and consistent communication regarding all aspects of AI deployment and operation.

To maximize value, the advantages of agentic AI should be managed and distributed equitably. This demands pervasive inclusion and equitable benefit through inclusive design practices for AI systems, ensuring they are usable by and fair to diverse populations. It also requires accessible enablement and training opportunities for all employees, and the active solicitation and integration of diverse perspectives in the governance and ethical oversight of AI.

Finally, and critically, leaders must cultivate deep psychological safety for ethical dialogue. Employees must feel secure to raise concerns about AI systems potentially leading to biased or unfair outcomes, without any fear of reprisal or marginalization. This open dialogue is essential for identifying and rectifying ethical blind spots and ensuring AI serves human values.

Driving this profound cultural transformation requires far more than internal memos or occasional town halls. It demands active, visible, and sustained leadership intervention in shaping daily work routines, reinforcing new collaborative rituals, explicitly rewarding behaviors that align with an agentic culture centered on trust and fairness, and addressing resistance or anxieties with transparency, empathy, and demonstrable action. Culture is the agentic operating system; viewing its transformation as a core strategic prerequisite is essential for sustainable success.

Intentionally cultivate pervasive trust, empower employees to advocate for fairness and quality, foster organizational adaptability, provide operational clarity regarding AI's role, and ensure pervasive inclusion – these are non-negotiable prerequisites for human-AI synergy, not optional "soft" initiatives.

Evolving Roles and Capabilities: Designing the Agentic Workforce

Cultural transformation must proceed in lockstep with the deliberate evolution of organizational structure, individual roles, and requisite capabilities. The integration of AI agents fundamentally reshapes workflows, decision points, critical skill requirements, and, importantly, how an organization identifies, develops, and leverages talent. Leaders must proactively engage in strategic workforce design, anticipating these shifts, identifying emerging needs, and systematically fostering new fluencies and competencies aligned with a high-performing Agentic Enterprise grounded in these principles. This is not merely an HR function; it is a core strategic imperative demanding C-suite attention and investment.

A key shift is to move from functional silos to fluid, fair, and inclusive collaboration. Rigid organizational silos and entrenched hierarchies constrain agentic potential and can perpetuate hidden biases. In the Agentic Enterprise, value creation often occurs through dynamic, cross-functional collaboration orchestrated by human-agent teams. Leaders must champion adaptable, outcome-oriented team structures, rethinking traditional reporting lines and empowering cross-functional collaboration that is explicitly designed to be inclusive and to leverage diverse perspectives. This fluidity should enhance visibility of talent and ideas across the organization.

The agentic shift also necessitates identifying and cultivating emerging strategic roles that act as stewards for these principles. These new roles bridge the gap between technology, operations, ethics, and strategy, with a strong focus on ensuring fairness and effectiveness. The Agent Experience Designer (AXD), for instance, designs intuitive, trustworthy, and fair human-agent interactions. They ensure AI interfaces are accessible, transparent about AI reasoning (to an appropriate degree), and do not inadvertently create barriers or biases for users. Their work is critical for ensuring AI tools are perceived as equitable partners. Another vital role is the Human-Agent Collaboration Lead / Fair Process Integrator, who designs protocols and processes for effective and equitable human-agent teaming. This individual coaches teams on leveraging AI tools fairly, optimizing blended workflows for both efficiency and alignment with fairness principles, and ensuring that AI recommendations are critically evaluated before action. Furthermore, the Autonomy & Ethics Program Manager / Fairness & Quality Assurance Lead defines operational and ethical boundaries for AI agents, aligns agent autonomy with risk appetite and regulatory requirements, manages validation processes (including bias and fairness testing), and ensures fairness principles are embedded and audited in agentic systems. This role acts as a key steward of Truly Fair and Relevant Evaluation in AI.

Concurrently, core functions must evolve. Operations Managers will shift from managing human-only teams to designing, optimizing, and overseeing human-agent workflows, with a keen eye on efficiency, ethical implications, and the fair distribution of tasks and developmental opportunities. IT/AI Platform Teams will manage agent platforms, infrastructure, and data governance, ensuring these systems support capabilities like bias detection tools and data representativeness to ensure fair and accurate outcomes. Human Resources / Talent Management pivots to strategic workforce planning that actively supports these goals. This includes redesigning talent acquisition to leverage AI for broader visibility (while mitigating AI bias), developing AI-augmented employee development pathways offering broad opportunity, and creating competency models and fluency training that incorporate awareness and ethical AI use. Risk & Compliance Functions must expand oversight to algorithmic decision-making risks, including systemic bias and fairness failures, partnering with business and technology for proactive mitigation. Finally, Product Owners / Service Designers will incorporate agent capabilities into roadmaps focusing on value and fair, ethical user experiences, measuring AI's impact on efficiency and fairness outcomes. This evolution demands strategic investment in upskilling and reskilling.

Crucially, leaders must cultivate essential agent fluencies across the organization, which extend beyond basic technical literacy. Collaboration Fluency is the ability to effectively delegate to AI agents, understand their capabilities and limitations, and work synergistically in hybrid teams, ensuring equitable contribution. Oversight Fluency entails critically evaluating AI outputs, identifying potential errors or biases, knowing when to trust recommendations, and when to intervene for Fair and Relevant Evaluation. Framing Fluency is the competence to translate business problems into clear, unambiguous goals for agents that are aligned with principles of fairness and effectiveness. A new, critical fluency is Fairness Advocacy: the understanding and ability to critically assess human-agent systems, identify potential for bias or inequitable outcomes, and effectively advocate for enhancements that improve fairness and performance.

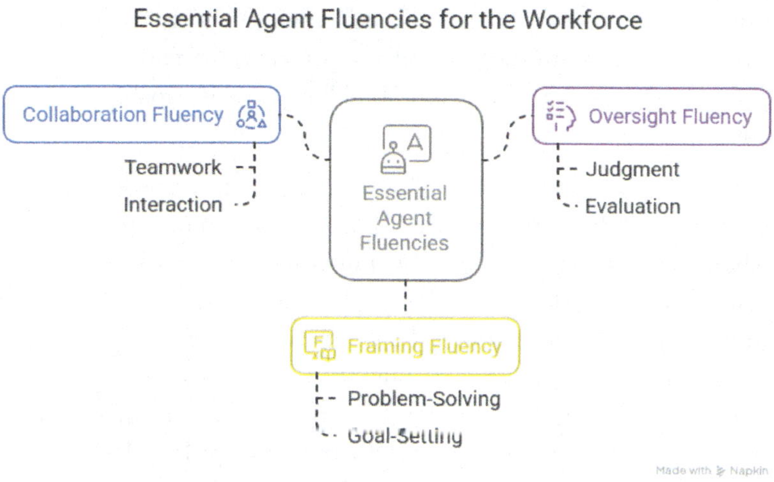

Figure 5-3. Essential Agent Fluencies (Fairness Advocacy not shown)

Systematically developing these fluencies is an ongoing organizational commitment, not a one-time event. Treating strategic workforce design as a core imperative – identifying roles, guiding evolution, and cultivating deep agent fluencies including fairness advocacy – is fundamental to unlocking the sustainable advantage of the Agentic Enterprise.

Proactively evolve roles, champion new fairness and quality stewardship functions, and systematically build "agent fluencies" (collaboration, oversight, framing, and fairness advocacy) across the organization to unlock potential and ensure fairness.

Building Agent Fluency: From Adoption to Advantage

Redesigning roles and identifying new capabilities sets the stage; realizing sustained value from agentic AI requires deliberate, continuous, and strategically targeted investment in building broad-based agent fluency. This extends far beyond basic "how-to" training. It encompasses a deeper operational understanding of how AI agents learn and reason, the critical thinking skills needed to interact with autonomous systems effectively and ethically, and the collaborative competencies essential for maximizing human-AI synergy effectively and fairly.

Deep fluency is a strategic imperative because surface-level understanding or uncritical acceptance of AI outputs leads to significant risks. These include underutilization of AI capabilities, over-reliance on potentially flawed or biased agent recommendations, inconsistent operational benefits, and costly errors with ethical and legal ramifications. True transformation requires a workforce that interacts with AI agents competently, critically, and with a constant awareness of the principles of fairness and visibility. This demands enablement programs focused on building genuine capability, sound judgment, and ethical discernment.

Leaders must champion and resource tailored enablement strategies, recognizing that a one-size-fits-all approach is insufficient. Training for Senior Leadership & Board Members should focus on strategic AI implications, enterprise risk management (including fairness-related risks), trustworthy AI governance, ethical leadership, and their role in championing a human platform built on trust and fairness. Managers & Team Leads need enablement on leading hybrid human-AI teams, fostering psychological safety for fairness advocacy, designing fair collaborative workflows, managing change, and coaching on ethical AI interaction. Frontline Employees & End-Users require practical training on effective AI tool interaction, critical evaluation of AI outputs (including bias identification), knowing when to exercise oversight or escalate concerns about bias or fairness, and relevant data literacy. Specialized & Support Functions (HR, IT, Legal, Risk, Audit, Procurement) need specific knowledge to support the agentic ecosystem; for example, HR must understand AI's role in fair talent processes, while Legal must navigate AI-specific challenges related to fairness and data privacy.

Building fluency over time benefits from a blend of continuous learning methods. Contextual Learning integrates training directly into workflows. Micro-Learning Bursts offer short, focused modules on specific AI concepts or fairness principles. Accessible Knowledge Hubs centralize best practices and resources. Communities of Practice facilitate peer-to-peer learning and problem-solving. Identifying and empowering Internal Champions & Fairness Advocates helps disseminate best practices. Lastly, Consistent

Communication & Reinforcement of successes, learnings, and the importance of these principles is vital.

Investing strategically in broad and deep agent fluency, infused with a strong understanding of fairness and visibility principles, is foundational for driving responsible AI adoption, mitigating multifaceted risks, maximizing AI ROI, and building a more capable, ethical, and successful organization.

Designing Human-Agent Collaboration Frameworks

Fluency equips the workforce, but the mechanisms governing their interaction with AI agents must be clear, efficient, and explicitly designed for fairness. Intentionally designing robust human-agent collaboration frameworks is a critical yet often under-addressed aspect of agentic transformation. Simply deploying AI agents and training humans on their use, without architecting how they will work together effectively and equitably, invites friction, errors, mistrust, and ultimately, failed synergy.

Realizing the promise of human insight amplified by machine consistency – and ensuring this partnership drives better outcomes by leveraging the full potential of the workforce – requires deliberate design choices. Leaders must ensure these frameworks address key dimensions of interaction, with fairness as a core design principle. This includes establishing an effective and fair handoff architecture. Seamless task and information flow between humans and AI agents is crucial to prevent bottlenecks and errors. Design clear protocols defining triggers for handoffs, the required context for each, and unambiguous completion criteria. Focus on logical intervention points that ensure fluid collaboration and, critically, that handoffs do not inadvertently introduce bias or disadvantage individuals. For example, tasks requiring nuanced human judgment should not be unfairly offloaded to AI, or vice-versa, in ways that impact developmental opportunities.

Reliable and impartial escalation pathways are essential for managing exceptions, errors, or ethically ambiguous situations safely and fairly. Define clear triggers for escalation, such as AI low confidence, detection of potential bias, user challenge to an AI decision, or concerns about fairness or bias. Specify routing protocols, information packaging requirements, and reviewer authorities, ensuring escalation is a documented, tested, managed, and impartial process providing a fair hearing for all concerns.

Collaboration is further enhanced by shared context mechanisms that promote fairness and effectiveness. Informed collaboration depends on both humans and AI agents having access to relevant, timely, and unbiased information. Architect secure agent access to core systems and data. Maintain comprehensive interaction histories (while rigorously respecting privacy regulations). Prioritize investments in data governance and knowledge management infrastructure that ensures data quality, representativeness, and minimizes the risk of agents operating on biased context, thereby promoting broad visibility and fairer evaluations.

Integrated feedback loops drive continuous improvement and fairness assurance. Transform human oversight from passive monitoring into active contribution to system improvement and ethical alignment. Implement intuitive interfaces for humans to provide feedback on AI performance, accuracy, usability, and perceived fairness or alignment with these principles. Establish clear processes to systematically collect, analyze, and act upon this feedback, ensuring resources are allocated to close the loop and make necessary adjustments.

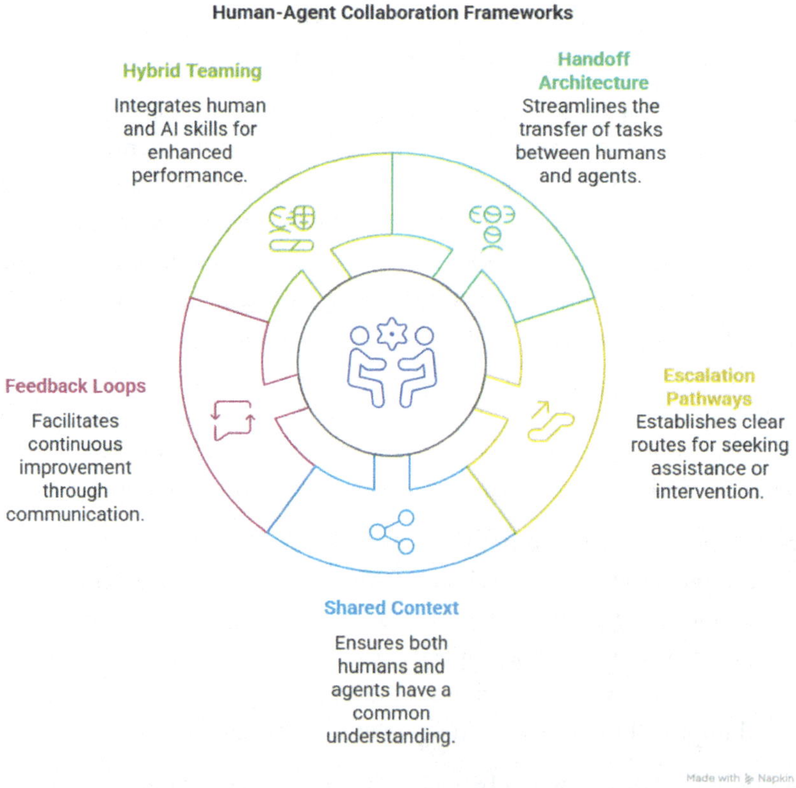

Figure 5-4. Human-Agent Collaboration Framework

Organizations should also consider hybrid teaming structures designed for high-performance synergy. Intentionally design roles and team structures to optimize human-AI synergy while promoting broad participation and fair workload distribution. Experiment with models where AI agents complement human expertise by handling routine tasks or initial analysis, freeing humans for complex problem-solving and ethical deliberation, as long as these structures elevate human contribution equitably.

Finally, a cornerstone of procedural fairness is designing for contestability, redress, and transparency. For AI-driven evaluations significantly impacting individuals (e.g., hiring, credit), frameworks must include clear mechanisms for individuals to understand the basis of the evaluation, challenge it if perceived as unfair, and have it reviewed by a human with authority to remediate.

Intentionally designing these collaboration frameworks, with fairness and effectiveness at their core, transforms human-agent interactions from ad-hoc encounters into reliable, scalable, ethical, and high-performance operational models.

Leadership's Role: Actively Architecting the Human Platform for an Agentic Future

Frameworks provide structure, and AI agents offer capabilities, but it is leadership that provides the vision, energy, resources, and unwavering direction. Nurturing, activating, and continuously refining the human platform is a fundamental, non-delegable, and intensely personal leadership responsibility in the Agentic Enterprise. Realizing the transformative potential of agentic AI while ensuring the organization operates on principles of fairness and unlocks its full potential demands active, visible, and sustained involvement from the highest levels of the organization. Your direct engagement is not optional; it is critical.

Leaders must champion the vision and foster trust through a commitment to fairness and transparency. This involves continuously articulating the strategic 'why' behind the agentic transformation, framing it not just as technological evolution but as an opportunity to build a more effective, innovative, and fair organization. Communication must be authentic and transparent, addressing potential benefits and challenges, including anxieties about job displacement or unfair AI decisions. Leaders should visibly model desired behaviors: curiosity, critical thinking about AI, and commitment to ethical conduct. Explicitly building and protecting psychological safety is paramount, creating an environment where employees feel empowered to question AI outputs, contribute to improving AI systems, and raise concerns about fairness or bias without fear of negative repercussions.

Driving cultural and structural alignment with these principles is another key leadership responsibility. This means actively shaping organizational norms to support human-AI collaboration grounded in fairness and dismantling silos that hinder diverse perspectives. Leaders should design and reinforce rituals celebrating ethical and effective AI use. Critically, performance management, incentives, and recognition programs must align to reward effective, ethical, and fair collaboration with AI agents, as well as efforts to improve fairness and quality in AI-driven processes.

Furthermore, leaders must invest in and empower talent for an AI-aware workforce. This requires championing sustained funding for tailored, continuous learning programs to build broad agent fluency, explicitly including principles of fairness and critical thinking. Active support for strategic role redesign elevating human capabilities and creating new fairness and quality stewardship roles is essential, as is ensuring equitable access to AI tools and training, creating broad opportunities for all.

Finally, leaders must ensure accountability for these principles and monitor progress. Establish clear metrics for tracking the human platform's health, including indicators that track fairness in AI-assisted processes and employee perceptions of fairness. Hold leaders at all levels accountable for cultivating a fair and high-performing human platform. Demanding demonstrable action on feedback regarding AI fairness reinforces a culture of continuous learning.

This direct, personal, and sustained leadership engagement distinguishes true, responsible transformation from superficial technological adoption. It is how you ensure your Agentic Enterprise is built on a foundation of human values and enduring trust.

Intentionally design frameworks for handoffs, escalations, shared (unbiased) context, and feedback to transform human-agent interaction into reliable, ethical, and high-performance models. Your personal leadership is the catalyst.

Case Study: Software Firm Unlocks Synergy via Culture and Focus on Fairness

Scenario: CodeCrafters Inc., a rapidly growing software development firm, deployed Collaborative AI Agents to augment developer tasks like code generation and bug detection. They anticipated immediate, significant productivity gains. Instead, adoption stalled. Developers expressed skepticism about the agents' reliability, fairness in how tasks were suggested or code evaluated by AI, and ambiguity in collaborative workflows. The technology existed, but the human dimension, particularly concerns around fairness and trust, was neglected.

Agentic Solution Connection: CodeCrafters' leadership realized success depended on cultivating the Human Platform with a strong emphasis on fairness and effective collaboration. They initiated an active cultural transformation. First, they championed the vision, clearly communicating that AI agents were to augment developer creativity, not replace them or create unfair evaluation systems, emphasizing fair and transparent AI use. Second, they invested in agent fluency (including principles of fairness and ethical evaluation) with tailored training on effective and ethical collaboration. This included recognizing potential AI biases and critically evaluating AI outputs for Fair and Relevant Evaluation. Third, they designed human-agent collaboration frameworks that promoted fairness and transparency, creating clear protocols for handoffs, feedback channels for agent performance and perceived fairness, and transparent processes for integrating AI-generated work. This addressed workflow ambiguity and built trust. Finally, they redesigned roles and emphasized human skills, clarifying that complex problem-solving and innovation remained core human responsibilities, with AI as powerful assistants. Senior developers were empowered as "Fairness and Quality Stewards."

This intentional, leadership-driven focus on the human platform, infused with these principles of fairness and transparency, transformed the AI agents from sources of skepticism into effective and trusted partners. Developer adoption soared, productivity gains were realized, and innovation quality improved as developers felt more empowered and fairly supported.

Key Questions for Leaders

- How are we strategically redesigning roles and workflows to deliberately leverage and amplify uniquely human capabilities (judgment, creativity, empathy, ethical oversight) in partnership with AI agents, while ensuring these partnerships are effective and fair?
- What concrete actions are we taking, beyond communication, to cultivate the essential cultural attributes—such as pervasive trust, psychological safety, and organizational adaptability—required for successful human-agent collaboration?
- Do we have a clear strategy and dedicated resources for building broad 'agent fluency' (collaboration, oversight, framing, and fairness advocacy) across the organization, tailored to different roles and responsibilities?

- Are we intentionally designing the specific frameworks for human-agent collaboration (handoffs, escalations, shared context, feedback loops, teaming structures, contestability of AI evaluations) needed for seamless, productive, and fair integration?
- How are we ensuring that our leadership team is personally modeling the behaviors and commitments necessary to build a fair and high-performing human platform in the agentic era?

Key Leadership Imperatives

Effective leadership in the Agentic Enterprise requires a steadfast commitment to several key imperatives for nurturing the human platform.

1. **Prioritize the Human Platform as the Cornerstone of Trusted AI.** Recognize that your people, culture, and collaboration models are not secondary concerns but the central foundation upon which successful, ethical, and effective agentic transformation is built. Allocate resources, attention, and strategic priority accordingly.
2. **Actively Architect Culture and Collaboration for Fairness and Effectiveness.** Intentionally shape cultural norms that foster trust, empower fairness advocacy, and build adaptability. Design explicit frameworks for seamless and equitable human-agent interaction, rather than leaving these critical elements to chance or allowing existing biases to be encoded.
3. **Invest Strategically in Talent Evolution and Agent Fluency.** Commit to continuous, targeted investment in building comprehensive agent fluency – including a deep understanding of fairness principles and ethical AI use – across the entire workforce. Proactively redesign roles to leverage unique human capabilities amplified by AI in a fair and empowering manner.
4. **Lead the Human Transformation Personally and Ethically.** Champion the vision of a fair and effective agentic future. Foster psychological safety for open dialogue about AI and fairness. Visibly model desired behaviors and ethical decision-making. Maintain unwavering accountability for nurturing the human platform and ensuring AI serves human values – this responsibility cannot be fully delegated.

These imperatives aim to capture the essential leadership actions required to successfully and ethically navigate the human dimension of the transition to an Agentic Enterprise, ensuring that technology serves to elevate human potential and fairness.

Conclusion: The Human Platform as Your Enduring Strategic and Ethical Advantage

The agentic journey, with its promise of unprecedented technological capability, hinges ultimately on the vitality, adaptability, ethical grounding, and cultivated capabilities of its human core. While AI agents provide powerful leverage, it is the strategic development of the human platform – one that is built on the principles of fairness and unlocking human potential – that unlocks sustainable value, fosters genuine innovation, and builds lasting differentiation.

Success in this new era demands active, visible, and non-delegable leadership focused on fostering pervasive trust, guiding the ethical evolution of roles, building deep and critical agent fluencies (including fairness advocacy), designing effective and fair human-agent interactions, and empowering the entire organization to navigate this transformation with confidence and integrity. Building the human platform is not an ancillary task; it is a central strategic and ethical imperative. Technology provides potential; the engaged, fluent, ethically grounded, and strategically directed human element translates that potential into breakthrough innovation, superior stakeholder relationships, adaptive organizational resilience, and enduring market leadership.

The most successful and respected organizations of the agentic age will be defined not just by the sophistication of their AI, but by the wisdom, creativity, ethical commitment, and strategic judgment of their people, intentionally elevated and empowered alongside their digital counterparts, within a system designed to be fair and to create broad opportunities.

Chapter 6: Sustaining Strategic Alignment & Future-Proofing

Beyond Initial Wins: Architecting Enduring Agentic Advantage

Successful AI agent implementation, robust trust frameworks, and a cultivated human platform establish your operational foundation. These are necessary and significant achievements, hard-won milestones on the path to becoming an Agentic Enterprise. However, initial wins, no matter how impressive, guarantee little amidst the relentless pace of technological evolution and the dynamic nature of competition. True leadership in the Agentic Enterprise extends far beyond the initial launch; it requires architecting an organization that remains perpetually adaptive, strategically aligned, and consistently resilient. The Agentic Enterprise is not a static destination but a continuously evolving state, demanding ongoing executive stewardship, strategic foresight, and proactive, often courageous, adaptation.

This final chapter addresses the crucial 'how-to-sustain' imperative essential for long-term leadership in this new era. We will examine the near-term trajectory of AI technology, acknowledging both its potential and its complexities. We will outline core principles for future-proofing your organization, emphasizing that resilience is built, not assumed. We introduce practical assessment frameworks for maintaining course, recognizing that navigation in this landscape requires constant vigilance. Finally, we reinforce the essential leadership mindset required for enduring success – one that embraces ambiguity, champions continuous learning, and remains steadfast in its commitment to ethical and value-driven AI. The objective is to equip you not merely to build an Agentic Enterprise, but to lead one that continuously learns, adapts, and maintains its competitive edge in a state of constant, intelligent motion.

The Shifting Horizon: Anticipating the Next Evolution of Agency

Sustaining any advantage necessitates clear, yet pragmatic, foresight into the evolving capabilities of AI agents. While today's agents deliver substantial value, leaders must anticipate powerful trends already poised to further reshape operating models and competitive landscapes, bringing with them both new opportunities and new governance challenges.

A key evolution is the decisive shift from discrete task automation to comprehensive outcome orchestration. The transformative opportunity lies not in simply automating individual steps more efficiently, but in designing interconnected agent systems capable of managing complex, end-to-end business outcomes. Imagine, for example, an AI system autonomously resolving a customer service issue – from initial detection and diagnosis, through cross-departmental coordination and resource allocation, to final, personalized communication and follow-up – all with minimal human intervention for standard execution. Mastering this level of orchestration requires a clear leadership mandate for ambitious, cross-functional process redesign, sustained investment in seamless, intelligent integration, and, critically, the development of sophisticated governance mechanisms capable of overseeing complex, emergent behaviors. The payoff is a potential step-change improvement in efficiency and effectiveness that far surpasses simple task optimization, yet the journey demands meticulous planning and rigorous risk management.

Facilitating this sophisticated orchestration often involves the rise of composite agents and dynamic agentic workflows. This entails deploying specialized AI agents, each optimized for specific functions (e.g., data ingestion, risk analysis, communication generation), that collaborate dynamically within larger, more complex processes. These "digital teams" can tackle far more intricate problems than standalone agents. However, they also introduce significant leadership challenges in system architecture, defining clear delegation boundaries and accountability for each specialized agent, managing the intricate governance complexity inherent in such interconnected and potentially interacting autonomous systems, and ensuring overall workflow integrity and resilience. Effectively architecting, governing, and managing these digital teams is rapidly becoming a required core competency for leading organizations.

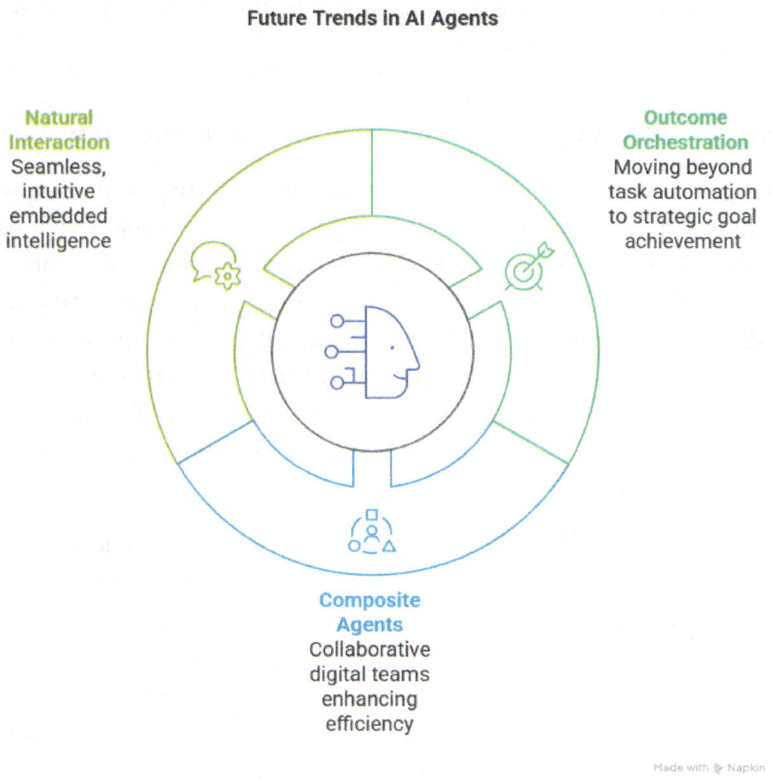

Figure 6-1. Future Agentic AI Trends

Simultaneously, human-agent interaction is moving towards more natural and ambient models. Leaders should anticipate agents with richer multimodal understanding – capable of processing and responding to voice, images, and complex data streams with greater nuance and contextual awareness. We will see more sophisticated intent-driven collaboration, where users state desired outcomes rather than issuing explicit, step-by-step commands. Pervasive ambient assistance, seamlessly embedded within daily tools and offering proactive, context-aware support, will also become more common. While these advancements promise lower adoption barriers and enhanced productivity, leaders must proactively govern the design and deployment of this embedded intelligence. This governance is crucial to mitigate risks such as user frustration from poorly timed or intrusive interruptions, unintended autonomous 'overreach' by proactive agents, ambiguity regarding control and authority, and significant data privacy implications arising from continuous environmental sensing. Anticipating and strategically preparing for these interconnected trends – outcome orchestration, composite agents, and natural interaction – is essential for maintaining a competitive edge and ensuring these powerful capabilities are harnessed responsibly and ethically.

Sustaining Strategic Alignment: Keeping Agents Tethered to Value

Anticipating the technological horizon is crucial, but ensuring that your deployed AI agents remain tightly tethered to current strategic priorities is equally vital for sustained success. As organizational goals shift and market dynamics evolve, AI agents – even those initially well-aligned – can quickly become misaligned. Such misalignment leads to wasted resources, diluted focus, a drag on agility, and potentially, agents optimizing for outdated or even counterproductive objectives. Sustaining strategic alignment is therefore not a one-time setup task but a continuous, dynamic leadership discipline, essential for maximizing the return on your AI investments and ensuring agents consistently contribute to core business objectives.

Failure to actively manage this alignment leads to significant value leakage and strategic drag. Leaders must embed standard operating procedures for constant evaluation and readjustment of agentic initiatives. This requires mandating strategic portfolio management for all AI agents, driven by regular, rigorous reviews – quarterly, for instance – involving key leaders from business, technology, strategy, and finance. These reviews must leverage a comprehensive Agent Registry (as discussed in Chapter 4) to systematically assess each agent's ongoing contribution to strategic goals, its current risk profile (including ethical and compliance considerations), its operational costs, and its continued alignment with enterprise priorities. This data-informed process, which must be more than a cursory check-in, should directly inform decisions on iteration, enhancement, resource reallocation, or, importantly, the decisive retirement of agents or initiatives no longer delivering strategic value or fitting the evolving strategic landscape.

Crucially, leaders must enforce outcome-centric performance measurement. The focus must shift from purely operational metrics ("Is the agent working correctly? Is it processing X transactions per hour?") to strategic contribution ("Is the agent delivering demonstrable strategic value? Is it advancing our core objectives? Is it enhancing customer outcomes or competitive positioning?"). This ensures that technical performance always serves a clear, measurable business purpose. Finally, leaders must drive agile prioritization and resource allocation, acknowledging that this can involve difficult choices. This involves implementing processes to quickly adjust agent goals, reallocate resources to higher-impact initiatives, or make tough but necessary de-investment decisions to retire low-value or misaligned agentic systems. This active, ongoing stewardship, which requires courage to stop initiatives as well as start them, ensures that your AI investments remain focused, deliver compounding value, and adapt alongside your evolving strategy.

Future-Proofing Your Agentic Enterprise: Principles for Enduring Resilience

Strategic alignment serves today's needs; enduring advantage, however, requires building inherent organizational resilience and adaptability. Future-proofing your Agentic Enterprise involves proactively architecting an organization capable of not just weathering continuous change, but of thriving within it, anticipating shifts, and even shaping them. This means embedding a set of reinforcing principles across the enterprise fabric, recognizing that future-proofing is an active, ongoing process, not a one-time fix.

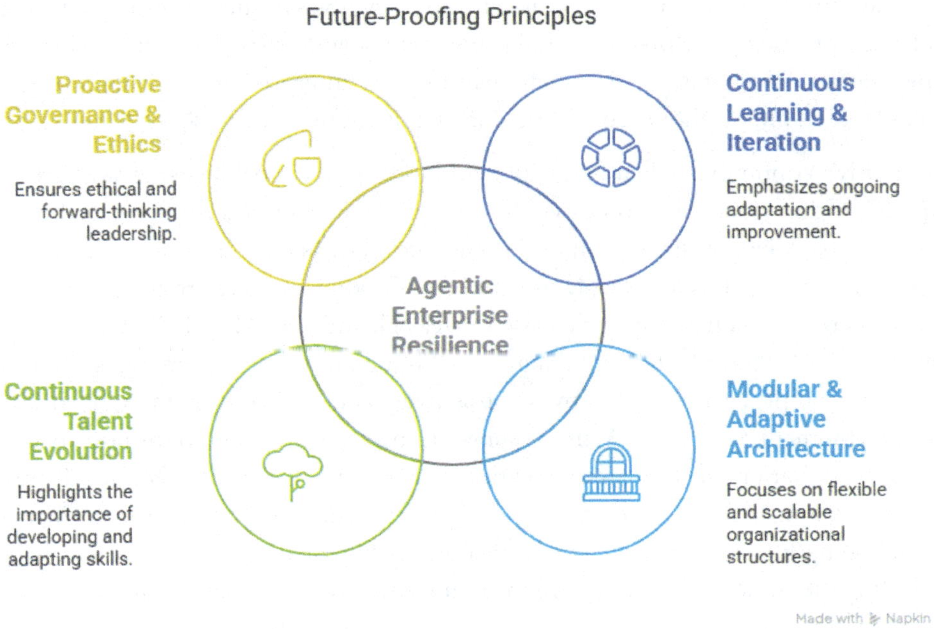

Figure 6-2. Future Proofing Principles

First, leaders must cultivate continuous learning and iteration as a core organizational discipline. Rigid organizations, resistant to change and new information, will inevitably be outmaneuvered in the rapidly evolving agentic landscape. Champion an adaptive learning culture where feedback loops – from customers, employees, and AI systems themselves – are actively used to rethink workflows, identify areas for improvement, and refine AI performance. Support safe-to-fail experimentation within clear ethical boundaries and implement agile review processes that enable rapid adjustments based on new insights and changing conditions. The executive mandate is clear: demand agility, foster the psychological safety necessary for honest feedback and learning from mistakes, and ensure your processes and culture reward evolution and adaptation.

Second, architect your technology for modularity and adaptability. Avoid brittle, monolithic AI platforms that create vendor lock-in, accumulate technical debt, and hinder future flexibility. Prioritize designs that feature reusable components, standardized APIs, and open architectures, allowing for easier updates, integrations, and the adoption of new AI advancements as they emerge. Guide technology investments towards platforms that explicitly prioritize adaptability, interoperability, and minimize deep, irreversible dependencies. This strategic approach ensures your enterprise remains ready for future technological shifts without requiring complete, disruptive overhauls.

Third, drive continuous talent and skill evolution. As AI capabilities advance, the skills required within your workforce will shift constantly and often unpredictably. Treat workforce upskilling, reskilling, and role adaptation not as episodic training events but as ongoing strategic priorities, deeply integrated with your business strategy. Continuously assess emerging skill needs – particularly in areas like human-AI collaboration, ethical AI oversight, data literacy, critical thinking in an AI-augmented environment, and the ability to govern complex AI systems. Invest proactively and substantially in targeted learning and development programs. Actively support employees as they transition to evolved roles where their unique human capabilities are augmented, not replaced, by AI, ensuring these transitions are managed equitably and transparently (as emphasized in Chapter 5). Championing lifelong learning signals unequivocally that people development is key to navigating the future successfully and maintaining a vibrant human platform.

Fourth, maintain proactive governance and ethical foresight. Regulatory frameworks and societal expectations surrounding AI are dynamic and will continue to evolve, often at a pace that challenges traditional compliance cycles. Reactive compliance is insufficient and inherently risky. Ensure your AI governance policies and ethical guidelines are treated as living documents, subject to regular review and adaptation in light of new technologies, emerging risks, evolving legal and societal norms, and learnings from your own AI deployments. Regularly revisit risk assessments for your AI systems, incorporating new threat vectors and potential societal impacts. Conduct periodic "ethical red teaming" to proactively identify potential vulnerabilities or unintended ethical consequences. Stay informed on evolving regulations globally, and adjust your internal policies preemptively, rather than waiting for enforcement actions. The leadership imperative here is to ensure your governance framework itself remains adaptive, positioning responsible and ethical AI as an enduring organizational commitment and a source of stakeholder trust. Future-proofing, in essence, is about deliberately building the mindset, flexibility, capabilities, and adaptive governance needed to navigate unforeseen changes successfully and responsibly.

"In 2024, U.S. federal agencies introduced 59 AI-related regulations – more than double the number in 2023 – and issued by twice as many agencies." – The State of AI 2025. This underscores the accelerating pace of regulatory change leaders must anticipate and proactively address.

Essential Disciplines for Continuous Command and Adaptation

Principles guide action; sustained navigation of the agentic future demands active steering through disciplined, recurring practices embedded in the leadership rhythm of your organization. Implementing potent assessment routines is essential for maintaining situational awareness, enabling informed course correction, and ensuring continuous alignment and adaptation. These are not bureaucratic exercises but vital instruments of strategic leadership.

Leaders must mandate continuous agent portfolio assessment. Deployed AI agents are not static assets; they are evolving capabilities and, potentially, evolving risks. Passive treatment or benign neglect is unacceptable. Ensure that your organization's resources are consistently generating maximum strategic value by instituting rigorous, scheduled portfolio reviews for all significant AI agents. This leadership discipline, supported by objective data and clear metrics (both financial and strategic), yields actionable intelligence for decisive resource allocation, targeted investment in high-performing agents, and proactive risk management or responsible retirement of underperforming or misaligned systems.

Leverage regular maturity model diagnostics to assess your overall agentic capability maturity holistically. This involves evaluating progress across key dimensions such as Strategy (clarity and alignment), Technology (scalability, adaptability), Governance (effectiveness, ethics, risk management), Workforce (skills, fluency, culture), and Measurement (value realization, outcome tracking). Such diagnostics help prevent organizational blind spots, identify areas of lagging capability, and guide balanced, strategic investment in areas requiring further development. Mandate periodic, candid self-assessments involving cross-functional leadership teams to create a data-driven, actionable roadmap for enterprise-wide agentic capability development.

Employ strategic scenario planning, augmented by AI agents themselves. Test your organization's strategic resilience against plausible future disruptions, technological shifts, or competitive moves. Integrate agent-augmented analysis into your regular strategic planning rhythms. Task specialized AI agents or dedicated analytical teams to model "what-if" scenarios, such as the impact of new competitor offerings leveraging advanced AI, sudden market changes driven by geopolitical events, or significant regulatory shifts impacting AI deployment. This practice sharpens organizational foresight, challenges existing assumptions, and informs proactive, anticipatory decision-making.

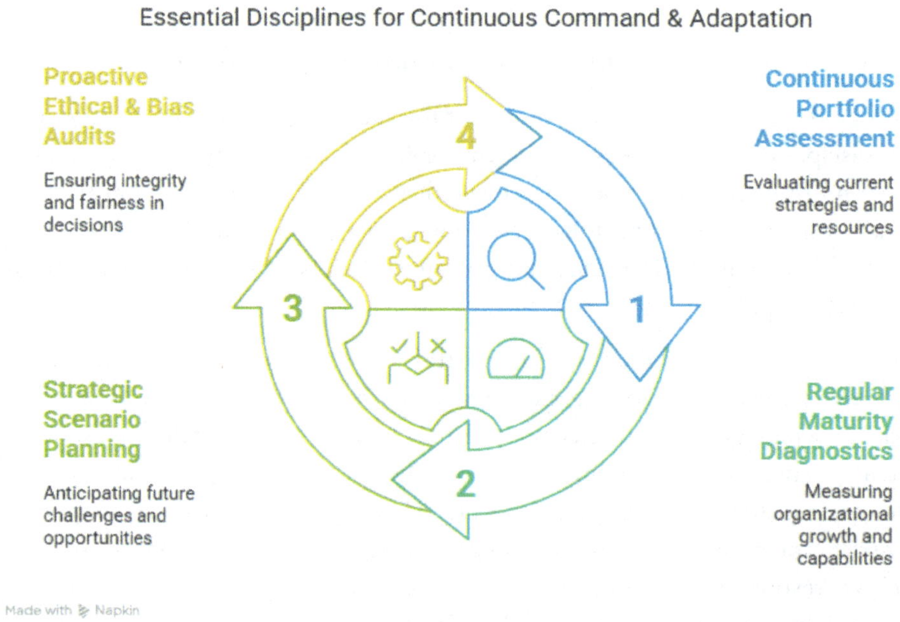

Figure 6-3. Continuous Command and Adaptation

Institute proactive ethical and bias audits for your AI systems, moving beyond mere compliance to genuine assurance. Trust, once earned, requires ongoing vigilance to maintain. Ensure your AI agents consistently operate within defined ethical values and do not perpetuate or amplify harmful biases. Establish recurring ethical and fairness audits for high-impact agents, reviewing their decision logs, training data, operational outputs, and stress-testing their behavior under various conditions, particularly looking for disparate impacts on different stakeholder groups. This provides ongoing assurance of integrity, reduces reputational risk, and reinforces your commitment to responsible AI and the VOF principles discussed in Chapter 5.

Implementing these assessment disciplines, and acting decisively on their findings, establishes the vital rhythm of strategic learning and adaptation required for effective command, control, and continuous improvement in your Agentic Enterprise.

The Leadership Imperative: Vision, Agility, and Enduring Commitment

Principles, frameworks, and assessment tools are insufficient without the quality and unwavering commitment of leadership. Sustaining strategic alignment and fostering organizational adaptability in the agentic era demands unwavering strategic vision, profound operational agility, courageous and data-informed decision-making, and steadfast personal resilience from every leader. Leaders must actively, and personally, steward the continuous evolution of their Agentic Enterprise. This is not a responsibility that can be fully delegated.

Leaders must maintain visionary steadfastness amidst turbulence. Focus relentlessly on your organization's fundamental strategic purpose and long-term value creation. Ensure all agentic initiatives clearly serve this defined North Star and avoid distracting, low-value technological pursuits or chasing fleeting trends. Continuously articulate this vision and filter all significant actions and investments through this strategic lens, providing clarity and stability for the organization.

Leaders must also actively architect organizational agility. Ensure your organizational structures, operational processes, talent development strategies, and cultural norms are designed to enable, not hinder, rapid adaptation and innovation. Champion continuous learning and foster psychological safety, empowering teams to experiment responsibly (within ethical bounds) and learn from both successes and setbacks. Dismantle bureaucratic hurdles that impede iteration and responsiveness. Empower teams with appropriate levels of autonomy within clear strategic and ethical guardrails. Ensure your governance framework itself remains agile, risk-calibrated, and adaptive to changing needs.

Exercising decisive resource stewardship is equally critical. Make courageous, data-informed judgments when allocating scarce resources (capital, talent, leadership attention), directing them to initiatives with the highest strategic potential and alignment.

Be equally decisive, and often more challenging, in de-investing from or retiring agentic systems or projects that are no longer serving core objectives, that present unacceptable levels of risk, or whose ongoing costs outweigh their strategic benefits. Leverage rigorous portfolio assessment for clear-eyed prioritization and actively champion the responsible retirement of outdated or misaligned systems to free up resources for future innovation.

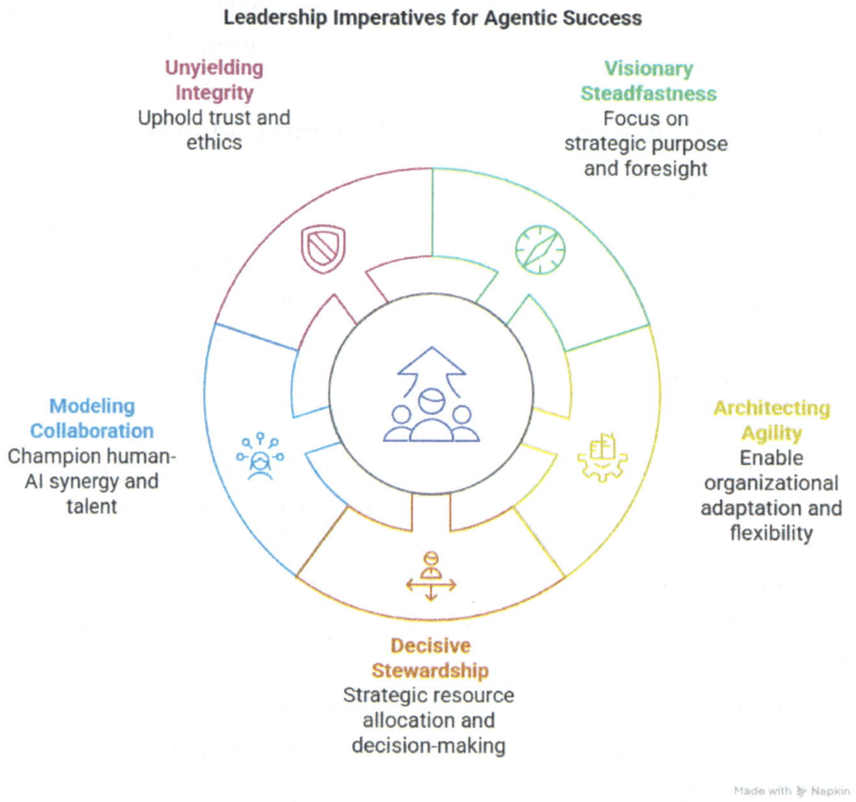

Figure 6-4. Leading the Agentic Future

Furthermore, leaders should model collaborative intelligence and human-centricity. Personally demonstrate effective, ethical, and critical engagement with agentic tools and AI-generated insights. Show, by example, how to leverage AI to augment human capabilities, not merely to replace them. Champion human capability development through comprehensive fluency programs and strategic role redesign, consistently reinforcing the unique and amplified value of human judgment, creativity, empathy, and ethical reasoning in an AI-augmented environment.

Finally, leaders must ensure unyielding integrity and trust stewardship. As the ultimate steward of your organization's values, reputation, and societal license to operate, guarantee that the pursuit of innovation never compromises ethical principles or stakeholder trust. Continuously reinforce ethical guardrails for AI. Ensure governance mechanisms adapt proactively to new challenges and evolving societal expectations. Demand transparency in AI operations and decision-making. Intervene swiftly and decisively when deviations from ethical standards or trust principles occur, and communicate openly and honestly, both internally and externally, about your organization's commitment to responsible AI use and its ongoing efforts to uphold that commitment.

Leading the evolution of your Agentic Enterprise requires this sustained, engaged, and principled leadership, focused on achieving both high performance and unwavering integrity. It is a demanding role, requiring both strategic acumen and profound ethical responsibility.

Case Study: Logistics Leader Maintains Edge Through Adaptation

Scenario: Global-Ship Logistics, having achieved initial success with AI agents in dispatch and route optimization, recognized the critical need to ensure these gains were sustainable and that they were prepared for the next evolution of agency. Complacency was not an option in their fast-moving and highly competitive industry, where new entrants constantly challenged established players.

Agentic Solution Connection: Leadership at Global-Ship instilled disciplines for continuous command and adaptation as a core part of their operating rhythm. They mandated continuous strategic alignment of their AI agent portfolio via rigorous assessments using outcome-centric metrics, ensuring every agent contributed directly and measurably to key business objectives like cost reduction, delivery speed, and customer satisfaction. They embedded future-proofing principles into their operational DNA: architecting their AI systems for modularity to facilitate easier upgrades and integration of new technologies, driving continuous talent evolution to ensure their workforce possessed the latest analytical and AI collaboration skills, and maintaining proactive, adaptive governance to navigate the changing regulatory and risk landscape for autonomous logistics. Agent-augmented scenario planning became a regular practice, helping them anticipate and prepare for shifts towards composite agents for complex logistics coordination and more sophisticated AI-driven outcome orchestration across their supply chain. This unwavering commitment to enduring resilience and proactive adaptation ensured their Agentic Enterprise remained a consistent source of competitive advantage and market leadership.

Key Questions for Leaders

- How will we proactively prepare for the next evolution of agency (e.g., Outcome Orchestration, Composite Agents, Natural Interaction), including the governance and human capability shifts required, to maintain our competitive edge, rather than merely reacting to changes as they occur?
- What disciplined processes (e.g., regular agent portfolio reviews with clear ROI and strategic value metrics, outcome-centric performance dashboards, agile resource reallocation mechanisms) will we embed in our leadership rhythm to ensure our agent portfolio remains continuously aligned with shifting strategic priorities and delivers maximum, measurable value?
- Which of the core future-proofing principles (Continuous Learning & Iteration embedded in culture, Modular & Adaptive Technology Architecture, Continuous Talent & Skill Evolution for an AI-augmented workforce, Proactive & Adaptive Governance for AI ethics and risk) requires the most significant investment, cultural shift, or fundamental change within our current operating model?
- Are we, as a leadership team, fully committed to implementing and sustaining the essential assessment disciplines (Agent Portfolio Assessment with clear de-investment triggers, Agentic Capability Maturity Diagnostics with actionable improvement plans, Strategic Scenario Planning integrated into strategy cycles, Proactive Ethical & Bias Audits with transparent reporting and remediation) needed for ongoing command, control, and adaptation of our Agentic Enterprise?

Key Leadership Imperatives

To ensure the Agentic Enterprise not only succeeds initially but thrives and maintains its advantage over the long term, leaders must embrace several critical imperatives.

1. **Mandate Continuous Strategic Alignment.** Actively and rigorously manage the AI agent portfolio against current business objectives using outcome-centric metrics and agile resource allocation to prevent value leakage and ensure focused,

high-impact contributions. This includes the courage to sunset underperforming or misaligned initiatives.

2. **Architect for Enduring Resilience.** Proactively embed future-proofing principles – a culture of continuous learning and responsible experimentation, modular and adaptive technology choices, ongoing talent evolution and upskilling, and dynamic, forward-looking governance – into the organizational DNA to build an enterprise capable of thriving amidst constant change and unforeseen disruptions.
3. **Instill Disciplined Assessment Routines.** Implement regular agent portfolio reviews, agentic capability maturity diagnostics, agent-augmented strategic scenario planning, and proactive ethical and bias audits as essential, non-negotiable instruments of command, control, organizational learning, and continuous adaptation.
4. **Lead with Vision, Agility, and Unwavering Integrity.** Maintain steadfast strategic focus amidst turbulence and complexity. Champion organizational agility and a culture of continuous improvement and responsible innovation. Make decisive and courageous resource choices based on data and strategic value. Ensure unwavering ethical stewardship and a commitment to trustworthy AI throughout the entirety of the agentic transformation and its ongoing evolution.

These imperatives summarize the key actions leaders must consistently take to guide their organizations effectively and responsibly through the evolving agentic landscape, transforming challenges into opportunities for sustained leadership.

Conclusion: Leading the Enduring Agentic Enterprise

Ultimately, sustaining advantage in the Agentic Enterprise is not a project with a defined endpoint, but an ongoing journey of strategic alignment, adaptive resilience, and unwavering leadership commitment. The ascent of agentic AI presents both profound opportunities for value creation and complex challenges in governance, ethics, and organizational adaptation. Mastering the disciplines outlined in this chapter – anticipating the shifting technological horizon with a realistic understanding of its implications, rigorously maintaining strategic alignment in a dynamic environment, proactively future-proofing the organization against unforeseen changes, and embedding robust routines for continuous command and adaptation – ensures your organization is positioned not just to implement agentic capabilities, but to lead enduringly and responsibly in the future of work. This commitment to perpetual evolution, guided by clear vision, strong principles, and courageous leadership, is the hallmark of true success in the agentic age.

Conclusion: Leading the Agentic Future

This guide has charted the emergence of the Agentic Enterprise – not as a distant forecast, but as a defining transformation unfolding in real-time. We stand at a pivotal juncture where the leadership choices made today will distinguish organizations that adapt, innovate, and thrive from those reshaped or, ultimately, displaced by the currents of change. The rapid ascent of capable AI agents fundamentally challenges every leader to rethink the nature of work, the architecture of their organization, and the very sources of sustainable competitive advantage.

This journey into the agentic era, however, is not solely a technological endeavor. Its success will be profoundly shaped by the quality of leadership, the resilience of organizational culture, and the intentional design of ethical, effective human-machine collaboration. Navigating this complex terrain successfully demands leading with strategic intent, grounded in unwavering core principles. It requires an unyielding anchor on business value, ensuring every agentic initiative serves core objectives rather than pursuing technological novelty for its own sake. It compels us to architect for true human-agent synergy, designing systems that amplify unique human capabilities – judgment, creativity, empathy, ethical reasoning – through intelligent and respectful collaboration.

Critically, this transformation mandates that we embed Trusted AI by Design™ from inception. This is not a mere compliance checkbox but the very bedrock of sustainable agentic success. This commitment makes proactive governance, robust ethics, comprehensive security, and operational transparency foundational, non-negotiable requirements. Success also means a commitment to scale capability, not just technology, moving deliberately through reusable patterns, adaptive governance, and fostering continuous organizational learning.

And perhaps most profoundly, realizing the full promise of the Agentic Enterprise calls upon leaders to spearhead the human transformation. This involves directly addressing workforce anxieties, investing strategically in new fluencies and skills, and personally modeling the collaborative, ethical, and adaptive future. It means championing a culture that unlocks the full potential of the workforce. This is achieved by embedding practical principles into the human platform: ensuring universal visibility so all potential is seen, creating equitable opportunities for participation, and applying fair and relevant criteria to all evaluations. Only by weaving these principles into the organization's fabric can leaders ensure that AI elevates all talent and serves all stakeholders justly.

The future belongs to organizations that embrace this moment with strategic ambition – guided by clear purpose, grounded in engineered trust, committed to fostering systems that unlock human potential, and dedicated to elevating that potential alongside intelligent automation. The Agentic Enterprise, when led with wisdom, courage, and integrity, represents a powerful canvas for building smarter, faster, more resilient, and ultimately more human-centric organizations. It is an opportunity to redefine not just how we work, but how we create value ethically and equitably for all stakeholders.

Your Leadership Imperative: A Call to Decisive, Principled Action

Understanding the contours of this new landscape is merely the prelude. True competitive advantage, and indeed responsible stewardship, flows directly from decisive, informed, and principled action. The strategic terrain is being redrawn now. Waiting is no longer a strategy; it is an active concession in a rapidly evolving world. Your leadership imperative demands immediate, intelligent traction:

1. **Ask the Crucial Questions – Now:** Convene your leadership team. Rigorously interrogate your organization's readiness. Challenge existing assumptions and explore the unique opportunities presented by agentic AI, using the frameworks and inquiries provided throughout this guide. Determine your immediate strategic priorities, your non-negotiable ethical guardrails, and your foundational commitment to Trusted AI by Design™ and to building a fair and high-performing human platform.

2. **Identify and Resource a Catalyst Pilot:** Move swiftly beyond discussion to tangible action. Task a focused, empowered, and cross-functional team to launch

a high-impact catalyst pilot designed to deliver measurable business value and critical organizational learning within a defined timeframe, such as 90 days. Ensure this pilot is not only a technical test but also an early proving ground for your principles of fairness and your trust-building efforts. Provide unwavering executive sponsorship and the necessary resources for success.

3. **Mandate Trusted AI by Design™ from Day One:** Insist that all AI agent deployments embed transparency, accountability, security, privacy, and ethical considerations – the core tenets of trustworthy AI – into their very architecture and operational lifecycle from inception. Establish minimum viable governance and rigorous validation processes, including bias and fairness assessments, *before* any launch that impacts stakeholders or core operations.
4. **Champion Fluency and a Culture of Fairness:** Visibly sponsor and invest in comprehensive enablement programs that build AI literacy, critical thinking, and fairness advocacy across the workforce. Communicate the 'why' behind the agentic transformation relentlessly and transparently. Address concerns with empathy and facts, and actively shape a culture that embraces collaboration, learning, ethical responsibility, and a deep commitment to fairness.
5. **Commit to Continuous Learning and Adaptation:** Treat initial deployments and strategies as first steps in an ongoing journey. Establish robust feedback loops, regular portfolio reviews, capability maturity assessments, and proactive ethical and fairness audits. These ensure your agentic capabilities remain aligned with strategic goals, ethical commitments, and the evolving technological and societal landscape.

This is the demanding, essential work required to position your organization to actively shape the agentic future, not merely react to it. The future of work, of value creation, and of organizational intelligence is being architected today. Lead boldly, lead wisely, and lead with unwavering integrity, building an Agentic Enterprise that is as trustworthy and fair as it is powerful and innovative.

About Bryant Research

The journey to becoming an Agentic Enterprise is the defining strategic challenge of our time. It demands more than technological implementation; it requires the courage to reimagine operations, the wisdom to govern with foresight, and an unwavering commitment to elevating human potential. Bryant Research partners with leaders to navigate this complex transformation, moving beyond the hype to architect real, sustainable value.

Our approach is built on the core principles of this guide. We focus on pragmatic strategy development that anchors every agentic initiative to clear business outcomes. We work with you to architect and embed *Trusted AI by Design*™, ensuring that proactive governance, robust ethics, and comprehensive security serve as the bedrock for confident scaling. Critically, we guide you in cultivating the Human Platform – the culture, skills, and collaborative models essential for ensuring that technology serves and amplifies your people justly and effectively.

Our expertise lies in synthesizing these domains: translating algorithmic possibility into board-level strategy, fortified by governance structures that manage risk and build trust, and fostering the cultural adaptations that unlock true human-agent synergy. For leaders ready to move from understanding to decisive action – whether that means launching a high-impact Catalyst Pilot, architecting genuinely trustworthy systems, or developing a comprehensive roadmap for an agentic future – Bryant Research serves as an experienced guide and implementation partner. We focus on delivering tangible results while building your internal capability, empowering you to lead your organization's transformation with clarity and confidence.

Contacting Bryant Research

To learn more about Bryant Research or explore how we can support your Agentic Enterprise journey, please visit BryantResearch.AI or contact us at BecomeAClient@BryantResearch.AI.

Glossary

Key Terms

Agent Fluency: The capability across the workforce to interact confidently, critically, and collaboratively with AI agents. It encompasses Collaboration Fluency (effective delegation and teamwork), Oversight Fluency (judging outputs and knowing when to intervene to ensure fairness and quality), Framing Fluency (translating business problems into clear goals for agents that align with fairness principles), and Fairness Advocacy (the understanding and ability to critically assess human-agent systems, identify potential for biased or inequitable outcomes, and advocate for enhancements that improve performance and ethical alignment).

Agent Registry: A centralized, dynamic inventory mapping every significant AI agent deployed or under development across the enterprise, documenting key details like ownership, purpose, risk tier, data sources, and governance status. It serves as a cornerstone for visibility, strategic portfolio management, enterprise risk oversight, compliance tracking, and ensuring clear lifecycle accountability for every agent.

Agentic Enterprise: An organization strategically re-architected to leverage AI agents as core components of its operating model. This transformation involves integrating autonomous systems into core processes for advanced decision-making, new forms of value creation, and deeply synergistic human-machine collaboration.

Agentic Scaling Models: Strategic organizational structures for deploying and managing AI agents as capabilities grow and diffuse throughout the enterprise. Models include a Centralized Model (offering maximum control and consistency, often suited for early stages or high regulation), a Federated Model (empowering domain-specific business units to develop and deploy agents within central standards and governance, enabling speed and local expertise), and a Hybrid Model (balancing central core infrastructure, governance, and expertise with empowered business units adapting and extending capabilities, often recommended for flexibility and control).

AI Agent / Intelligent Agent: An AI system engineered to perceive its environment (through data, sensors, or user inputs), reason through complex information, learn from interactions, plan multi-step actions, and autonomously execute tasks (often including the dynamic use of digital tools and systems) to achieve specific, often complex, goals with a significant degree of delegated authority. This distinguishes it from simpler automation like RPA or basic chatbots.

Autonomy Engine (AI Agent Component): A sophisticated core component of an AI agent responsible for interpreting high-level strategic goals and autonomously initiating, managing, prioritizing, and adapting complex sequences of actions without requiring constant, step-by-step human instruction or intervention for every decision. It enables the agent to plan and act independently within defined operational and ethical boundaries.

AXD (Agent Experience Designer): An emerging strategic role focused on designing intuitive, trustworthy, fair, ethical, and explainable human-agent interactions. The AXD ensures that AI interfaces are accessible, transparent about AI reasoning (to an appropriate degree), and align with user needs, brand standards, and principles of fairness and effectiveness, mitigating bias and enhancing usability.

Build/Buy/Partner Strategy: A strategic decision framework leaders use for acquiring or developing agentic capabilities. This involves evaluating whether to primarily **Build** solutions in-house (maximizing control and IP, for core differentiators), **Buy** off-the-shelf vendor products (prioritizing speed and simplicity for standardized needs), or **Partner** with external experts (blending speed, expertise, and capability building). A "Flexible Stack" often combines these approaches.

Burning Platform: A strategic situation characterized by intensifying operational, competitive, and stakeholder pressures that create an urgent, almost existential, imperative for an organization to undertake significant and rapid change, such as adopting agentic AI. Delay or inaction in such a scenario actively leaks value, erodes competitive positioning, and confers compounding strategic disadvantage.

Catalyst Pilot: An intense, strategically focused, and time-bound (typically 60-90 days) initial AI agent deployment. Its purpose is to unequivocally prove tangible business value quickly on a P&L-visible problem, generate critical organizational learning and momentum, build stakeholder confidence, and de-risk broader investment. It follows a disciplined, three-phase Plan-Prove-Scale framework: Plan (focused alignment, strategic de-risking, and defining Minimum Viable Governance), Prove (developing and deploying a Minimum Viable Agent to demonstrate measurable results), and Scale (leveraging success, codifying learnings into a reusable enterprise playbook, and planning for broader strategic impact).

CCPA (California Consumer Privacy Act): A key United States (California-specific) data privacy regulation that, alongside others like GDPR and HIPAA, governs the handling of personal data and presents critical considerations for data governance and compliance when deploying AI agents.

Center of Excellence (CoE) for AI: An organizational structure (which can be virtual or formal and dedicated) that serves as a central hub for cultivating and disseminating AI expertise, best practices, design standards, reusable tools, and proven methodologies. It facilitates knowledge sharing, capability building, and the consistent application of Trusted AI by Design™ principles and governance frameworks, thereby enabling and accelerating responsible, strategically aligned agentic innovation across the enterprise.

Collaborative Agent: An AI agent archetype designed as a sophisticated digital teammate to work interactively and synergistically alongside human experts on complex, ambiguous, often creative or strategic tasks. Examples include co-authoring reports, supporting intricate design processes, or providing intelligent decision support by synthesizing research and modeling scenarios, thereby augmenting and amplifying human cognition rather than automating it away.

Composite Agents: An advanced agentic architecture where multiple specialized AI agents, each optimized for specific functions (e.g., data ingestion, risk analysis, communication generation, ethical review), collaborate dynamically within larger, interconnected workflows. These "digital teams" can tackle far more intricate problems and orchestrate more complex outcomes than standalone agents but require sophisticated governance for their emergent behaviors and interdependencies.

Conversational Agent: An AI agent archetype specializing in natural, context-aware language interactions with users (such as customers, employees, or partners). Leveraging sophisticated memory, reasoning, and often LLMs, they can handle complex inquiries, provide personalized guidance, facilitate transactions, resolve issues, and maintain brand consistency at scale, moving beyond simple FAQ chatbots.

Future-Proofing: A proactive and continuous leadership discipline of architecting an organization – encompassing its strategy, technology, processes, talent, culture, and governance – for enduring resilience and adaptability. This enables the enterprise to not only navigate but also thrive amidst continuous technological evolution, market shifts, and unforeseen disruptions, rather than simply predicting specific future events.

GDPR (General Data Protection Regulation): A key European Union (EU) data privacy and protection regulation that imposes strict rules on collecting, processing, storing, and managing personal data. It is highly relevant to how AI agents handle personal data of EU residents, with significant implications for agent design, data governance, and compliance.

Governance (in context of AI Agents): The comprehensive framework of policies, ethical guidelines, standards, processes, controls, risk management protocols, and clear accountabilities. This framework is designed to ensure that AI agents are developed, deployed, and operated reliably, securely, ethically, transparently, fairly, and compliantly, remaining aligned with strategic objectives and societal values. It is particularly crucial given the autonomy, learning capabilities, and potential impact of AI agents.

Hallucinations (AI): Instances where AI models, particularly Large Language Models (LLMs), generate outputs that are plausible-sounding but factually incorrect, nonsensical, irrelevant, or entirely untethered from the provided input data or established knowledge. This poses a significant risk to the accuracy, reliability, and trustworthiness of AI agent communications and decisions.

HIPAA (Health Insurance Portability and Accountability Act): A key United States federal law that governs the privacy and security of Protected Health Information (PHI). It is highly relevant for AI agents that access, process, or store healthcare-related data, requiring stringent data handling, security measures, and compliance protocols.

Human-in-the-Loop (HITL) / Human-Over-the-Loop (HOOL): Design approaches for human-agent collaboration. **HITL** involves specific steps or decisions within an agent-driven process that mandatorily require human review, approval, or intervention before proceeding (e.g., for high-stakes actions or low-confidence outputs). **HOOL** implies a more supervisory role, where humans monitor agent operations and can intervene if necessary, but the agent has more autonomy for routine execution. Both are key for safety, accountability, and managing complex or sensitive contexts.

Human Platform: The collective capabilities, skills, ethical grounding, collaborative norms, adaptability, and engagement level of an organization's workforce. It is viewed as the critical, dynamic foundation upon which the transformative potential of agentic technology is either fully realized or hindered. Cultivating the Human Platform involves a strategic commitment to fairness and creating equitable opportunities to ensure AI serves to unlock the entire workforce's potential.

Instruction Followers: A term describing earlier generations of automation tools, such as Robotic Process Automation (RPA) and basic scripted chatbots. These tools are designed to execute meticulously predefined, explicit steps or respond to specific prompts within relatively rigid, pre-programmed constraints, lacking the ability to dynamically plan or adapt to achieve broader goals. (Contrast with Outcome Pursuers).

Language Interface / LLM Layer (AI Agent Component): A core component of many advanced AI agents, typically powered by Large Language Models (LLMs) and other natural language processing (NLP) technologies. It enables the agent to understand natural language instructions (textual and spoken) with a high degree of nuance, contextual awareness, and intent recognition. It also allows the agent to generate fluent, coherent, contextually appropriate, and often persuasive responses and communications, serving as the primary bridge for human-agent and, increasingly, agent-to-agent interaction.

Lifecycle Management (Agent): A structured, disciplined, and continuous approach to stewarding AI agents throughout their entire existence. This typically encompasses six interconnected phases: Ideation (aligning with strategic intent and defining value), Design & Validation (engineering for trust, performance, and scalability), Deployment & Initial Monitoring (controlled operational integration and intensive oversight), Ongoing Monitoring & Maintenance (sustaining operational excellence and compliance), Iteration & Improvement (driving adaptive evolution and enhancing value), and Retirement & Decommissioning (responsible sunsetting and knowledge preservation). This ensures sustained value, operational resilience, risk management, and responsible stewardship.

LLM (Large Language Model): A type of foundational Artificial Intelligence model, often based on deep learning architectures like transformers, trained on vast and diverse datasets of text and code. LLMs are capable of understanding, generating, summarizing, translating, and predicting human-like language with remarkable fluency, often serving as the core language interface, reasoning engine, or knowledge base for sophisticated AI agents.

Maturity Model (Agentic Capability): A structured framework used for assessing an organization's overall progress, proficiency, and capability development across key dimensions essential for successfully becoming and sustaining a leading Agentic Enterprise. These dimensions typically include Strategy (clarity, alignment, vision), Technology & Data (scalability, adaptability, infrastructure, quality), Governance (effectiveness, ethics, risk management, trust), **People & Culture** (skills, agent fluency, collaborative culture, fairness), and Measurement & Value Realization (outcome tracking, ROI).

Memory & Context Management (AI Agent Component): A critical core component of an AI agent that enables it to retain, recall, synthesize, and intelligently apply relevant information across multiple interactions, diverse contexts, and over extended periods. This includes storing and utilizing individual user preferences, detailed interaction histories, complex organizational policies, evolving procedures, dynamic product knowledge, market conditions, and critical environmental context. Well-managed memory is crucial for deeply personalized stakeholder experiences, adaptive and consistent agent behavior, continuous learning, and the effective management of intricate, long-running processes, but requires stringent data governance.

Minimum Viable Agent (MVA): The simplest, yet functional, version of an AI agent developed during a Catalyst Pilot. It is designed to perform the core targeted function within a real (though potentially constrained) operational environment to generate tangible, measurable business results and actionable user feedback. The MVA prioritizes rapid value demonstration and learning over comprehensive feature completeness.

Minimum Viable Governance (MVG): The essential, initial set of governance principles, risk assessments, operational boundaries, human-in-the-loop (HITL) points, and preliminary privacy, security, and ethical review protocols established for a Catalyst Pilot. MVG is mandated from the absolute outset to ensure responsible innovation, build stakeholder confidence, and generate a credible, trustworthy, and replicable win, rather than treating governance as an afterthought.

Monitoring Agent: An AI agent archetype that functions as an intelligent sentinel or observer. It continuously analyzes vast streams of data, system logs, market signals, or other environmental inputs to proactively detect subtle anomalies, emerging trends, potential risks (e.g., fraud, cybersecurity threats), or nascent strategic opportunities that might be invisible to standard reporting or human oversight alone. They typically alert human operators or trigger pre-defined automated responses based on their findings.

Outcome Orchestration: A sophisticated capability of advanced AI agent systems involving the design and implementation of interconnected agents capable of autonomously managing complex, end-to-end business processes and achieving strategic outcomes (e.g., resolving a multifaceted customer service issue from initial detection and diagnosis through cross-departmental coordination to final, personalized communication and follow-up). This represents a significant leap beyond automating discrete tasks, requiring minimal human intervention for standard execution pathways.

Outcome Pursuers: A defining characteristic and fundamental evolutionary leap of AI agents, distinguishing them from prior generations of automation tools (Instruction Followers). AI agents are engineered as "outcome pursuers" that leverage a suite of integrated cognitive capabilities (reasoning, planning, memory, tool use) to understand high-level goals, and then autonomously determine, plan for, and execute the necessary actions to achieve specified, often complex, business outcomes, frequently navigating dynamic, unpredictable, and data-rich environments with intelligent adaptability.

PII (Personally Identifiable Information): Any data that can be used, on its own or in conjunction with other information, to identify a specific individual. Examples include name, address, social security number, email address, or biometric data. Handling of PII by AI agents requires stringent data governance policies, security measures, and compliance with privacy regulations (like GDPR, CCPA, HIPAA) due to significant privacy risks and legal obligations.

Pillars of Trustworthy Systems (Trusted AI by Design™): The core, interconnected, and non-negotiable principles that form the foundational DNA for building and maintaining stakeholder confidence in agentic AI systems, as defined by the Trusted AI by Design™ methodology. These pillars are: Transparency & Explainability (illuminating the "why" and "how"), Accountability & Ownership (establishing clear responsibility), Fairness & Bias Mitigation (designing for equity and preventing harm), Security & Privacy (protecting assets, data, and individuals), and Safety & Reliability (ensuring predictable, dependable, and safe operation).

Prompt Injection: A significant and novel security vulnerability specific to AI agents, particularly those powered by Large Language Models (LLMs). Attackers craft malicious inputs disguised as legitimate prompts, instructions, or data to manipulate the agent's behavior, potentially causing it to bypass safety protocols, reveal confidential information, execute unauthorized actions, generate inappropriate content, or perform other unintended operations.

RBAC (Role-Based Access Controls): A fundamental security mechanism ensuring that users or systems, including AI agents, are granted access only to the specific information, functionalities, and resources that are strictly necessary for their explicitly assigned roles and defined tasks. This adheres rigorously to the principle of least privilege, minimizing potential damage from errors, misuse, or security breaches.

Reasoning Module (AI Agent Component): A core cognitive component of an AI agent that endows it with the sophisticated ability to perform complex mental tasks essential for intelligent action. These include planning detailed and adaptive task sequences to achieve strategic goals, evaluating multifaceted trade-offs between different potential courses of action (often under uncertainty or with incomplete information), handling ambiguity with greater efficacy, applying intricate business rules or complex regulatory policies in context, and dynamically adapting plans and behaviors when encountering unexpected obstacles or new critical data. It is the heart of the agent's advanced problem-solving capability and its capacity for exercising "judgment" within its defined operational domain.

RPA (Robotic Process Automation): A technology that uses software "robots" or "bots" to mimic repetitive human actions when interacting with the user interfaces of digital systems and software applications. RPA is typically employed for automating highly structured, explicitly defined, and rule-based tasks, such as data entry or simple data reconciliation, and is often contrasted with the more adaptive, cognitive, and goal-oriented capabilities of AI agents.

Strategic Workforce Design: A core C-suite strategic imperative, distinct from routine HR functions, that involves proactively anticipating and planning for the evolution of organizational structures, individual roles, critical skill requirements, and talent development strategies resulting from the deep integration of AI agents. This includes identifying emerging capability needs (such as 'agent fluencies' and fairness/quality stewardship roles), fostering new competencies through targeted upskilling and reskilling, and systematically reshaping the workforce to maximize human-AI synergy within a fair and high-performing Agentic Enterprise.

Task Agent: An AI agent archetype highly specialized for executing well-defined, often linear, and typically repetitive processes or discrete tasks with high precision, speed, and reliability within clearly established parameters. Examples include automated data extraction from documents, sending standardized notifications, or managing basic data entry. While valuable for tactical efficiency and streamlining specific operational choke points, they possess limited reasoning capabilities and are generally unsuitable for tasks requiring significant judgment or dynamic adaptation to novel situations.

"Three Cs" Framework (Pilot Selection): A practical diagnostic lens and decision-making framework for evaluating and selecting optimal candidates for initial AI agent deployments or Catalyst Pilots. A strong candidate use case must convincingly satisfy all three criteria: **Clarity** (the business problem is thoroughly understood, its strategic importance is clear, the agent's goal is specific, measurable, and directly tied to a key business outcome, and success metrics are unambiguous and agreed upon), **Control** (the agent's operational scope and decision boundaries can be clearly defined, reliably constrained, effectively monitored, and robust mechanisms for human oversight, intervention, and rapid correction are firmly in place), and **Confidence** (key stakeholders—users, process owners, and leadership—realistically trust the agent's outputs, the integrity of its data sources, the rigor of its oversight processes, and have high collective confidence in successful execution, value delivery, and the ability to manage potential risks effectively).

Tool Use (by Agents) / Tool Use & Integration (AI Agent Component): A critical capability and core component of advanced AI agents that enables them to dynamically select, securely access, and intelligently interact with a wide array of other digital tools, enterprise systems (e.g., CRM, ERP), operational technologies, external data sources, APIs, databases, or even other specialized AI services. Through sophisticated tool use, an AI agent translates its internal "decisions" and "plans" into tangible business operations, measurable real-world impact, and cross-functional orchestration to gather information or execute actions necessary to achieve its goals. Granting this capability requires stringent, uncompromising governance due to its high-value, high-risk nature.

Trusted AI by Design™: A strategic philosophy and comprehensive operational methodology that proactively and systematically embeds core principles of trust – including Transparency & Explainability, Accountability & Ownership, Fairness & Bias Mitigation, Security & Privacy, and Safety & Reliability – into the entire lifecycle of AI agent conception, design, development, data management, deployment, ongoing operation, and eventual retirement. This approach treats these critical elements as foundational architectural requirements and non-negotiable design imperatives, rather than as afterthoughts, separate compliance checks, or impediments to innovation, thereby enabling confident scaling, sustainable innovation, and enduring stakeholder confidence in the Agentic Enterprise.

Workflow Agent: An AI agent archetype designed to intelligently manage, coordinate, and orchestrate multi-step task sequences or end-to-end operational processes that often span interconnected systems, data sources, and human team members. They typically operate according to predefined (but often adaptable and complex) business logic, handling conditional branching, exception management, and seamless handoffs. Common use cases include employee onboarding, complex customer service journeys, or new product development lifecycles. Complexity/Risk is considered Medium.

Recommended Readings

Artificial Intelligence Index Report 2025

Stanford University Human-Centered Artificial Intelligence (HAI) (2025)

Provides comprehensive data and analysis across research, development, deployment, ethics, policy, economics, education, and public opinion regarding AI. Essential context for understanding the rapidly evolving landscape in which Agentic Enterprises operate and make strategic decisions.

State of AI Report 2024

Nathan Benaich (Air Street Capital) and collaborators (2024)

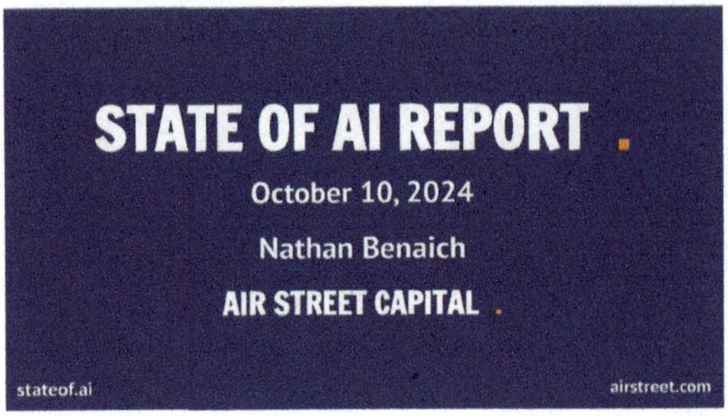

Offers insightful analysis focused on key breakthroughs and trends across AI research (including agents), industry dynamics (compute, economics, major players), safety considerations, and the shifting political and regulatory environment. Valuable for anticipating near-term shifts impacting strategy.

Index

Made in United States
Orlando, FL
18 June 2025

62207951R00083